Introduction to Technical Problem Solving with MATLAB

second edition

Jon Sticklen, PhD

Michigan State University
East Lansing, Michigan USA

M. Taner Eskil, PhD

Isik University
Istanbul, TURKEY

John L. Gruender

Editor

Great Lakes Press, Inc.
St. Louis, MO

PO Box 550 / Wildwood, MO 63040
(636) 273-6016
custserv@glpbooks.com
www.glpbooks.com

International Standard Book Number: 1-881018-37-7 (softcover)

All comments and inquiries should be addressed to:
Great Lakes Press, Inc.
c/o John Gruender, Editor
PO Box 550
Wildwood, MO 63040
jg@glpbooks.com
phone (636) 273-6016
fax (636) 273-6086

Library of Congress Control Number: 2006928761

Printed in the U.S.A.

10 9 8 7 6 5 4 3 2 1

This book belongs to:

_____

_____

_____

Contents

Chapter 7

Chapter 8

Acknowledgements

Michigan State University students enrolled in Computer Science and Engineering 131 (*Technical Problem Solving with Computer Tools*) have used portions of this book over the last several years. I am grateful to these students for their thoughtful feedback on the content in earlier versions, and for their lively discussions about "what works" for freshman students. I am also indebted to several colleagues at other universities who have given feedback on draft versions, most notably Prof. Howard Fulmer, Villanova University. Of course, no reviewer is responsible for any final content of this text. I am indebted to my colleagues Marilyn Amey, M. Taner Eskil, Timothy Hinds, and Mark Urban-Lurain, who formed a working group with me and were instrumental in developing the pedagogy underlying this textbook.

I wish to dedicate this book to a number of people. First, to my supportive and loving wife Peggy, I thank you for putting up with many chores gone undone and many weekends lacking normal family outings. Second, to my two wonderful daughters, Mitra and Cynthia, I thank you for putting up with long lapses of fatherly attention. Finally, to my extended family in Connecticut, thank you for your support at a distance, and especially to my father-in-law, Prof. Emeritus of Religion, Parker Lansdale. Thank you eternally, Parker, for teaching me that trying to understand students from their viewpoint can be frustrating and unproductive, but that in the end it's the only path for a "teacher."

Dr. Jon Sticklen, College of Engineering, Michigan State University
January 16, 2005

My learning experience in artificial intelligence, image processing, pattern recognition and many other technical fields through MATLAB started in the Pattern Analysis and Machine Vision Laboratory of Bogaziçi University (BUPAM). I am grateful to Prof. Aytül Erçil, the founder and the former director of the BUPAM laboratory, for the excellent learning environment she provided me and many other young researchers. I have also profited a great deal from my Ph.D. studies in the Computer Science and Engineering Department, Michigan State University, under the direction of Dr. Jon Sticklen (in computer science) and Dr. Clark Radcliffe (in mechanical engineering).

I would like to thank my parents, Hasan and Emine Eskil, for their continuous moral support and encouragement. None of my accomplishments would have been possible without their love and care.

Finally, I am eternally grateful to have my beloved Fezal Okur in my life, who has always inspired and challenged me and stood next to me in both my best and worst times.

Dr. M. Taner Eskil, College of Engineering, Michigan State University
January 16, 2005

Preface
(Primarily for Instructors)

This material below is intended primarily for instructors though students may find it of interest as well. In this preface, we will explain the role of this book, the target student audience, software versions of MATLAB® compatible with this book and instructor options for getting MATLAB in the hands of their students, learning objectives for this book, and the pedagogy backing this book.

I   Role for this Textbook and Target Student Audience

Over five hundred textbooks exist on MATLAB and its use in problem solving. So, the natural question is "Why another?" The reason is simple. Existing texts on MATLAB serve two audiences well: working professionals who, by necessity or desire, want to add MATLAB to their problem-solving toolset and advanced undergraduate and graduate students in formal university classes.

This text is written specifically to support independent learners and formal classes in MATLAB at the university freshman level.

Technical problem solving is a difficult topic for freshman university students. A specific stand-in for the general topic *technical problem solving* would be what freshman typically call *word problems*. Though nuts and bolts understanding of mathematical and physical system principles is usually sound in incoming freshman students, these same students have an aversion to applying their tool-knowledge to word problems in which the problem specification must be clarified, with the appropriate tool selected and applied.

Likewise, effective use of MATLAB as a tool for technical problem solving is difficult for typical freshmen students. Despite the many textbooks in print on MATLAB, most of these texts are written with a pedagogical perspective and a rate of presentation that overwhelms freshmen and leaves them with a sense that their "goal" is to memorize an impossible plethora of MATLAB built-in functions.

The problem does not lie in the impossibility of the goal; rather, the problem is that it's the wrong goal. Even seasoned MATLAB system developers do not know everything about MATLAB but are confident with a mind-set for problem solving that MATLAB enables and with a relatively small subset of MATLAB built-in functions. Further, the seasoned MATLAB programmers embody the abstract concept of life-long learner. They are well versed in using MATLAB Help to find MATLAB functionality that matches the problem currently under consideration. The top-level goal of this text is to provide a bridge between what typical freshman engineering and science students know and a working knowledge of a MATLAB kernel that over time a learner can expand on an individual, discipline-specific, as-needed basis.

This text is an outgrowth of six years of experience teaching a computer tools course (primarily MATLAB) targeted primarily at freshman students in the Michigan State University College of Engineering. The explicit purpose of this book is to meet the needs of freshmen as they learn the rudiments of MATLAB and apply their new tool-knowledge to solve technical problems. Though designed to support freshmen engineering courses, this text is appropriate for students from the natural sciences, some areas of business (such as accounting), social science disciplines, and students in any discipline in which quantitative methods are used as staples supporting problem solving.

For students to be effective in following the treatment followed in this text, some level of mathematical sophistication is prerequisite. The pre-requisite/co-requisite that demonstrates the appropriate level is "Calc 1" as it exists in most university class catalogs. The question of appropriate pre-req or co-req for the material in this book should be set on a local basis, however. In the best of worlds, students using this text would have solid understanding of high school algebra and trigonometry and would have completed high school classes in science through physics. This sort of background, however, is not recommended because material in this book builds on these subjects factually. Rather the recommendation is made because students who have (for example) a solid physics course under their belts are more likely to have internalized the idea of modeling physical systems with abstract symbol systems. Understanding abstract symbols and the manipulation of symbol systems is not required for learning with this text, but it makes learning easier from a student's perspective.

II Software Versions Assumed in This Book / Getting MATLAB Into the Hands of Your Students

Material in this book assumes that students will have available MATLAB 7 (Release 14). MATLAB 6 (Release 13) is in almost all ways compatible with material here, too. Earlier versions of MATLAB—5 and older—differ significantly from MATLAB 6 and MATLAB 7. Although this text could be used with versions MATLAB 5 and earlier, such use is not recommended.

From the perspective of a freshman learner, the default organization of the user interface is the main difference between MATLAB 6 and MATLAB 7. If your students are using MATLAB 6, some care is needed especially when students are assigned Chapter 3, Section 3-1 in which they bring up MATLAB for the first time. This section lays out what students will see when they launch MATLAB, and they discover the utility of the main MATLAB window. Since first-timers can be unsure of themselves when launching MATLAB, if your students are using this text and MATLAB 6, you should hand out to your students several screen shots showing the default organization of the main MATLAB window under MATLAB 6.

Using the book for instruction without the availability of MATLAB for student use will not work. The essence of learning to apply MATLAB for problem solving is to *use* MATLAB. You have two major ways of presenting MATLAB to your students.

The first is a Student Version of MATLAB. The Student Version is a personal version of MATLAB; your students will purchase the Student Version and install it on their personal computer. Caution your students to get the Student Version that matches their PC or Macintosh computer. The Student Version of MATLAB 7 is not a toy, nor is it a reduced version of the Professional Edition of MATLAB 7. Most of the core toolboxes that an undergraduate engineering student will need are available as add-ons to the Student Version. Indeed, the Student Version (no toolboxes additional) includes all that a freshman student (and well beyond) will need for a computational environment.

The Student Edition of MATLAB is available from The MathWorks at http://www.mathworks.com. In addition to direct sales from The MathWorks, your campus bookstore or computer store may have copies of the Student Edition. The cost of the Student Edition is about $100 retail. Tell your students the Student Version will be useful throughout their undergraduate experience, spreading the cost over four years as opposed to one term.

The second major way for you to present MATLAB is through an institutional installation of the MATLAB Professional Edition. If you have a public or scheduled lab associated with your course, your computer support organization can purchase a set of floating licences appropriate for your public lab setting. If you want to pursue this option, your computer support people should contact The MathWorks, or a local campus representative of The MathWorks.

Whether using the Student Edition or the floating license arrangement, you should ensure your students have the most recent available version of MATLAB. This is particularly important if you choose to have MATLAB stocked for sale in your campus bookstore or computer store. There are many software suppliers serving university bookstores and computer stores. In my experience, many of these suppliers will give a "good deal" on an older version of MATLAB. Unless its made clear to the manager of the bookstore or computer store which release of the Student Version should be stocked, your students are left open to potentially buying an out-of-date copy of MATLAB.

Complete contact information for The MathWorks is:

> The MathWorks, Inc.
> 3 Apple Hill Drive
> Natick, MA   01760-2098 USA
> Tel: 508-647-7000
> Fax: 508-647-7001
> Email: info@mathworks.com
> Web: www.mathworks.com

III   General Learning Objectives and Linkage to Textbook Content

Eight General Learning Objectives (GLO) operationalize the underlying learner goals that students will achieve after successful completion of the material in this text.

> **GLO #1:** **Students will be able to read an ill-structured, technical problem specification and transform it into a well-structured problem.**
>
> Chapter 2 (A Framework for Technical Problem Solving) focuses on a single example problem. The problem set is based on projectile trajectory, a standard physics problem. But the problem is set

and emphasizes the importance of problem refinement leading from an ill-structured problem to a well-structured problem specification and on the step-wise progression toward solution.

As students progress through the material of this book, the course instructor must emphasize the real-world importance of being able to read, understand, and transform an initial problem specification into a final one to be solved. My experience with freshman learners indicates that a good term project, defined in a suitably open-ended manner, can provide a "capping experience" which drives home the importance of this learning objective.

GLO #2:   **Students will be able to select an appropriate MATLAB tool (of those currently studied) to apply to a given problem.**

Chapter 3 through Chapter 8 contain two problem exercise sets at the end of each chapter. The first exercise set (called *Set A*) focuses on the mechanics of MATLAB *use* in the context of the given chapter. The second (called *Set B*) focuses on *selection* and *application* of the tools introduced in the chapter.

GLO #3:   **Students will become familiar with and be able to use the following components of the MATLAB standard user interface: command window, history window, work space window, current directory window, array editor window, and M-file editor window.**

Section 3-1 (The First Time You Open MATLAB) in Chapter 3 introduces the student to the MATLAB environment. Depending on student demographics, it may be essential for instructors to factor in the "newness" of the modern, complicated-looking user interface of the MATLAB computational environment – as perceived by freshman learners. The introduction in Section 3-1 is to give freshmen the big picture of the user interface as well as detailed working knowledge.

As the entire text unfolds, achieving this general learning objective is central.

GLO #4:   **Students will be able to create and debug MATLAB user-defined functions.**

Functions in computer science are an example of *abstraction* and *encapsulization*. More broadly, in science and engineering, effective system decomposition is dependent on abstraction and encapsulization. Students must understand and appreciate that user-defined functions are an underlying theme in this textbook.

User-defined functions are introduced in Chapter 4 (Saving Your Work in MATLAB) and are used in all later chapters, particularly in the exercise problem sets. Chapter 4 ends with an introduction to the debugging facilities in MATLAB.

GLO #5: **Students will be able to create valid scalar, vector, and array/matrix expressions and will apply their knowledge to solve freshman-level technical problems.**

Scalar expressions are introduced in Chapter 3 (MATLAB Basics: Scalars), vector expressions in Chapter 5 (Vector Operations), and array expressions including matrix operations in Chapter 7 (Arrays). These chapters, along with Chapter 8 (More Flexibility: Introduction to Conditional and Iterative Programming), represent the kernel of instruction presented for the essential MATLAB issues.

The order of these four chapters is significant and emphasizes targeting this text to freshmen. As described in more detail in the section below, the underlying pedagogy of this book follows the principle of introducing conceptually new material to freshmen by working and linking outward from familiar material.

GLO #6: **Students will be able to utilize the MATLAB programming constructs IF and FOR to create conditional and iteration-based, user-defined functions.**

The core MATLAB-as-tool material culminates with Chapter 8 (More Flexibility: Introduction to Conditional and Iterative Programming). This chapter focuses on the MATLAB **IF** construct and the MATLAB **FOR** construct. More advanced conditional constructs, notably **SWITCH**, are not introduced, to avoid conceptual overload. Likewise, the iterative construct **WHILE** is not introduced.

GLO #7: **Students will be able to utilize MATLAB 2-D plotting to create data charts that communicate well.**

Chapter 6 (2-D and 3-D Plotting and Using MATLAB Help) introduces making two-dimensional plots using MATLAB. Emphasis is placed on producing plots that communicate well, including insistence on chart labeling.

GLO #8:   **Students will be able to use MATLAB HELP to extend their ability to apply MATLAB.**

The ability of learners to bootstrap themselves from the material of this book to other capabilities in MATLAB is essential for all students who intend to use MATLAB as a tool for technical problem solving. Another topic in Chapter 6 (2-D and 3-D Plotting and Using MATLAB Help) focuses directly on learner use of the MATLAB HELP facility.

IV   Underlying Pedagogy for this Textbook

The following is part of a paper presented at the 2004 annual meeting of the American Society for Engineering Education (ASEE).[1] It is reproduced with the express consent of ASEE. The purpose of including the excerpt is to expose the pedagogy underlying this text in order to enable instructors to make better use of this text in their classrooms.

The work below was a team effort including our co-authors: Marilyn Amey (MSU College of Education), M. Taner Eskil (MSU College of Engineering), Timothy Hinds (MSU College of Engineering), and Mark Urban-Lurain (MSU College of Natural Science).

Application of Object-Centered Scaffolding to Introductory MATLAB Learning

Introduction

In this report, an extension to the standard sense of scaffolding as used in the learning literature is proposed. *Object Scaffolding* is proposed as a pedagogical technique in which current student learning is anchored by a conceptual map resultant from previous learning. This report focuses specifically on the use of object scaffolding for freshman-level instruction in MATLAB.

---

1. Sticklen, J., Amey, M., Eskil, T., Hinds, T., & Urban-Lurain, M. (2004, June 20-23). Application of object-centered scaffolding to introductory MATLAB. Paper presented at the American Society of Engineering Education Annual Conference & Exposition, Session Number 3553, Salt Lake City, Utah.

The conceptual starting point is the observation that freshmen engineering students do not arrive in introductory MATLAB courses possessing significant experience with array/matrix formulation of problem solutions, nor with computational operations such as matrix multiplication. Given this observation, rapid introduction of MATLAB starting with array/matrix operations from the outset is ill advised.

However, beginning engineering students do have background in scalar operations. Starting with "MATLAB as Scientific Calculator" and focusing on scalar expressions and becoming familiar with the MATLAB environment, then moving systematically to "Vector Operations," and finally to "Array and Matrix Operations" enables student linkages from the outset with past learning. Once student conceptual organization for MATLAB scalar operations solidifies, understanding of vector operations is a small step, with the most significant new concept being one-dimensional indexing into a vector. From there, two-dimensional indexing becomes a small step, and this opens the way for introduction of the full range of array and matrix operations that is linked to past student learning.

In this report, a description of a MATLAB learning progression, Scalars => Vectors => Arrays/Matrices, will be given, and the expected impact of introducing MATLAB by following this progression will be described.

Background—Concept of Scaffolding

As a concept, *scaffolding* is known in the engineering education community. But, as is often true across diverse technical areas, *scaffolding* can refer to different concepts depending on the commentator and the context for use.

When discussed in a pedagogical context, scaffolding is usually linked to the learning theories of Vygotsky (Vygotsky 1962; Vygotsky 1978). Vygotsky's viewpoints are rooted in the belief that a learner's cognitive development is enabled by interaction with more capable members of the same culture, usually teachers or other students. As students learn, they occasionally encounter difficulty, and at such times, appropriate assistance will help learners break through. The assistance can be of almost any form including the posing of Socratic-type questions to lead the learner, the revealing of unknown (to the learner) relevant facts that the learner needs to know, the description of apt analogies to help the learner, etc. The act of providing this assistance to the learner is termed *scaffolding*.

In Vygotsky's view, many times a learner will be on the verge of understanding a concept or solving a problem and will need a nudge from a more capable member of the same culture. That point of need for specific help is known as the state of *proximal development*. In the context of a formal class, the role of the teacher is to supply the needed assistance or to facilitate the learner getting the assistance from someone else. In Vygotsky's theory, scaffolding is undertaken only when a learner needs it, and, perhaps more important to the learning theory, that scaffolding is stopped as soon as possible, that is, as soon as the learner can solve exemplars of the target problem on her own.

A subtle point in the above paragraph often leads to confusion of what scaffolding (in the Vygotsky sense) means. Following Vygotsky, scaffolding is not an informational or procedural "fact" of stand alone character. Rather, scaffolding is the *act* of providing such facts that are currently needed by a learner but unknown to that learner. This follows from an implication of the importance in the Vygotsky theory that scaffolding be removed once the learner no longer requires it. A fact like "the MATLAB plot function has optional arguments for plot line color" is something that stays with a learner. The act of offering the nudge can be stopped once the learner is fully capable of independent work on target variety problems.

In the Vygotsky sense, the goal of scaffolding is that the learner gradually attains mastery in solving target variety problems. Moreover, such mastery should be demonstrable by the learner solving problems without help. In essence, the theory of proximal development is a road map for how to achieve this goal.

In the computer science community, *scaffolding* as described above has most often surfaced in the human factors subcommunity or in the intelligent tutors subcommunity. (Chalk 2001; Heffernan 1998; Jackson, Krajcik et al. 1998; Quintana, Krajcik et al. 2002) These subcommunities have taken the sense of scaffolding more or less intact from the original meaning delineated by Vygotsky.

In contrast, the term *scaffolding* has been used in the software engineering subcommunity in a more direct analogy to the scaffolding that a construction crew uses to erect a new building, that is, as an actual object used to build some final structure. (Sellink and Verhoef 2000) In software engineering, scaffolding has most often been used to mean software put in place whose sole purpose is to facilitate debugging. The high-level distinction between *scaffolding* as used in learning theory and *scaffolding* as used in software engineering is that in learning theory, scaffolding refers to *process*, while in software engineering, scaffolding refers to *object*.

Major Thesis of This Work

The major thesis of work reported here is that broadening the pedagogical usage of *scaffolding* to include the object sense (as used in software engineering) can shed light on learning difficulties faced by many engineering students when they are introduced to the MATLAB computational environment and by extension to other modern computational tools for technical problem solving.

If an educational scaffold is viewed as a conceptual map that a student must have to enable further student progress towards computational tool mastery, then a high-level shift in instructor viewpoint is needed. Scaffolding in the process sense remains an available pedagogical tool to facilitate targeted student learning for specific tool-use skills. But in the larger perspective of becoming familiar with and competent in the use of an entire computational environment, scaffolding in the object sense lays the groundwork for *systematic construction of a piece wise instructional presentation* of the computational environment in such a way that student learning is enhanced.

In a nutshell, the idea is that a current conceptual map is an enabling starting point for learner mastery of new items, and that the current conceptual map provides *anchors* for the learner as new topics are introduced. The current conceptual map then becomes the scaffold (in the object sense) for anchoring new student understanding.

Applying the Object Version of Scaffolding for MATLAB Instruction

To apply the object version of scaffolding, the first task is to decompose a broad learning objective into a number of identifiable stepping stones. To illustrate, set the broad learning goal to be the development of student facility in formulating and solving problems in an array/matrix format. In addition, a freshman engineering course focused on MATLAB use would extend to working with polynomials in MATLAB, to 2-D and 3-D graphing, to optimization problem solving, and so on. But for purposes of clarity of exposition here, we will limit the goal to a base line facility in using arrays and matrices.

Since freshmen typically have, at best, a theory-only comprehension of arrays and do typically have a solid working understanding of scalar operations, a first stepping stone would naturally lie in scalar computations using MATLAB. Students know how to use a calculator to solve such relationships as $D = \dfrac{2v^2 \sin(\phi)\cos(\phi)}{g}$ for the horizontal distance D traveled by a projectile fired with initial velocity v at an angle ϕ from the horizontal, given the acceleration due to gravity g. They do *not* initially know the scalar operator syntax of MATLAB, nor the concept of stored variables, nor a number of other facets required to perform scalar computations using MATLAB. But by isolating one educational target (scalar operations) and setting it up as one substantial unit of study, prior student knowledge of scalar computations becomes the conceptual anchor for students to leverage in assimilating new knowledge.

After the unit (stepping stone) on scalar computations in MATLAB is completed, including substantial exercise by learners on practice problems, vector operations becomes a small conceptual step. Vectors in MATLAB are linear, ordered sets of scalars. The scalar operations in MATLAB carry over directly to their vector analogues. For example, if **A** and **B** are two MATLAB vectors, the vector **C** = **A** + **B** is found by adding the corresponding elements in **A** and **B**. Thus, **C**(2) (the second element of the **C** vector) is found by **C**(2) = **A**(2) + **B**(2), and similarly for all other elements. In MATLAB, such operations are termed "cell-by-cell operations." The pedagogically important point is that following a unit on scalar operations with a well-circumscribed unit on vector operations builds cleanly on student understanding of scalar computations while introducing one new type of computation: cell-by-cell computations.

Table 1: Values for Total Horizontal Distance (initial speed = 50 feet per second)

ϕ (degrees)	D (feet)
0	0.0000
10	26.7203
20	50.2178
30	67.6582
40	76.9381
50	76.9381
60	67.6582
70	50.2178
80	26.7203
90	0.0000

A sample of the type of problem students are given in the unit on MATLAB vector computations is to revisit the equation first given as a scalar relationship for the horizontal distance traveled by a projectile. For a fixed value of the initial projectile speed and a known value for *g*, students can produce a table (Table 1) of values for *D* that result from a range of values for ϕ. ϕ is represented as a vector of values, and the same mathematical relationship is used but implemented by using vector operators instead of the scalar operators used before. The result is a vector of corresponding values of **D**. From the table results, students can approximately determine the firing angle that will *optimize* the horizontal distance the projectile covers.

As learners incorporate knowledge of vector operations into their conceptual map of MATLAB use, natural generalizations of the scalar operators are made to corresponding vector operators. The top-level knowledge to be learned is the concept of cell-by-cell operations on vectors. Once that is mastered the entire conceptual map for MATLAB use of vector computations can be understood as a generalization of the conceptual map for scalar computations.

So far, the learner has evolved two versions of a conceptual map for MATLAB use from a starting conceptual map for the use of a calculator to perform scalar computations. The first conceptual map, capturing student understanding of using MATLAB for scalar operations, depended strongly on the starting conceptual map. The second conceptual map, capturing student understanding of using MATLAB for vector operations, depended strongly on the first conceptual map.

Similarly, once student understanding of MATLAB use for vector operations is in place, the next stepping stone is array and vector computations in MATLAB. As in the case for student learning of MATLAB vector computations based on generalizing what had previously been learned on MATLAB scalar computations, student learning of MATLAB array/matrix computations is presented as a high-level generalization of vector operations. The progression in student learning is depicted in Figure 1 as the

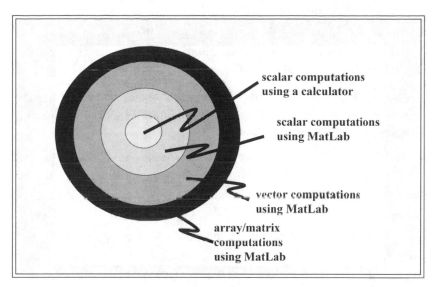

Figure 1: Layers of an Onion View of Evolving Conceptual Understanding of MATLAB

"layering of an onion."

It is important to emphasize that the view of evolving student learning of MATLAB in Figure 1 is predicated on well-circumscribed units on scalar computations in MATLAB, vector operations in MATLAB, and array/matrix computations in MATLAB. Without the circumscription of each stepping stone in the sequence, or to put it more operationally in terms of course layout, without each unit having a stand-alone character, systematic student generalization from what they know to what you intend for them to learn is impeded.

Conclusion

The proposal put forward addresses the need to ameliorate the steep learning curve of MATLAB as perceived by freshman engineering students. The proposal can be boiled down to the following:

- Decompose the initial major goal for MATLAB learning (pro-ficiency in scalar, vector, and array/matrix computations) into three self-contained units.

- In a first unit aimed at proficiency in scalar computations, make connections to student understanding of scalar compu-tations using a calculator.

- In a second unit aimed at proficiency in vector computations, portray vector computations as a generalization of scalar MATLAB computations.
- In a third unit aimed at proficiency in array/matrix computations, portray array/matrix computations as a generalization of vector MATLAB computations.

This recipe for MATLAB instructional design may appear to be straight forward and, in some ways, trivial. In fact, it is not self-evident as indicated by the most currently used MATLAB textbooks which blur the three units into one introductory chapter.

This report is a response to the steep student-perceived learning curve of freshman instruction in MATLAB. But the idea proposed to extend *scaffolding to support learning* to include *process scaffolding* (the "traditional sense" of scaffolding in learning theory) and *object scaffolding* (as proposed in this report) may find application beyond MATLAB instruction. This broadening for what the community calls scaffolding is prompted by one key observation: Students will learn most naturally when they have mastered a well-circumscribed set of concepts and have extended their understanding to new concepts in incremental steps.

Process scaffolding and object scaffolding can be viewed as folding in the concept of *proximal development*, as described by Vygotsky. When learners are in a state of proximal development, they are learning something new, something not understood, but close to being understood. In a process scaffolding setting, the small bit of missing information can be filled in by nudges from an instructor. The process of nudging is scaffolding. In an object scaffolding setting, the small distance between the new material (what the student knows) is bridged by a generalization of what the student has assimilated. In both process and object scaffolding, a small conceptual distance exists between what the student knows and what we hope the student will learn. The zone of proximal development is a common concept to both variants of scaffolding.

Process scaffolding is most appropriate for details (at the small grain). That is, process scaffolding is most applicable when a student is struggling with one relatively narrow facet of target learning. On the other hand, object scaffolding is aimed much more at the top-level layout of instruction. Process scaffolding and object scaffolding can be viewed as complementary learning enablers.

Process scaffolding and object scaffolding differ in one sense. With process scaffolding, the concept of *fading* is important. That is, once the learner is capable of problem solving without the nudge of process scaffolding, the scaffolding should be removed. But in object scaffolding, there is no fading or withdrawal of the scaffold.

Indeed, to the extent that object scaffolding is effective, the learner's conceptual map evolves and retains conceptual map knowledge from earlier stages of learning while integrating new understanding.

Though the view that scaffolding should be broadened to include both process and object scaffolding is well suited to act as pedagogical underpinning to support instructional design for freshman instruction in MATLAB, the concept of the two complementary types of scaffolding to support learning may find wider applicability.

Bibliography

Chalk, P. (2001). *Scaffolding Learning In Virtual Environments*. Annual conference on Innovation and technology in computer science education, Candelabra, UK.

Heffernan, N. T. (1998). *Intelligent Tutoring Systems Have Forgotten the Tutor: Adding a Cognitive Model of Human Tutors*. Conference on Human Factors in Computing Systems, Los Angles, CA, USA.

Jackson, S. L., J. Krajcik, et al. (1998). *The Design of Guided Learner-Adaptable Scaffolding In Interactive Learning Environments*. SIGCHI conference on Human factors in computing systems.

Palm, W. (2001). *Introduction to MATLAB 6 for Engineers*, McGraw Hill Higher Education.

Quintana, C., J. Krajcik, et al. (2002) *A Case Study to Distill Structural Scaffolding Guidelines for Scaffolded Software Environments*. SIGCHI conference on Human factors in computing systems: Changing our world, changing ourselves, Minneapolis, MN, USA.

Sellink, M. P. A. and C. Verhoef (2000). *Scaffolding for Software Renovation*. Fourth European Conference on Software Maintenance and Reengineering.

Vygotsky, L. S. (1962). *Thought and Language*. Cambridge, MA, MIT Press.

Vygotsky, L. S. (1978). *Mind in Society*. Cambridge, MA, Harvard University Press.

Wertsch, J. V. (1985). *Cultural, Communication, and Cognition: Vygotskian Perspectives*. Cambridge, UK, Cambridge University Press.

V   Topics to Be Included in the Sequel

A number of MATLAB tool and application topics are not covered in this textbook and will be covered in a forthcoming "sequel." Some of these topics could be considered core, yet they have been omitted because this text is specifically targeted for *beginning* engineering students. The topics and their presentation order in the current textbook were selected to be in accord with our pedagogical approach as described.

The most important fundamental topics to be covered in the sequel will focus on broadening the set of MATLAB representational tools. We will address MATLAB multi-dimensional arrays, strings and string arrays, cells and cell arrays, and structures and structure arrays. Though we have introduced strings, the treatments were informal and focused on the use of strings for labeling graphs. We will deal with strings separately and in depth. Multi-dimensional arrays, cell arrays, and structure arrays will also be covered in detail.

The central goal of the new chapters on additional representational tools will be two-fold. First, the nuts and bolts of the representational methods will be covered. Second, and more important from a use perspective, we will emphasize learner ability to match a given problem to an appropriate representation method.

In addition, several specific constructs will be introduced in the new text allowing more flexibility in building MATLAB programs. We will cover **WHILE** and the **SWITCH** conditional form. The current text included only **FOR** loops in the discussion of iterative constructs. The decision not to include **WHILE** loops in this initial book was largely due to experience with freshmen, which showed that freshmen can be challenged by loops in any form. Once the **FOR** construct is learned, moving on to **WHILE** constructs becomes simpler from a student perspective than learning both simultaneously.

The new book will also cover the basics and the application of **Simulink** and **State Flow**. **Simulink** is MATLAB's tool for building dynamic simulation models in a visual environment. **State Flow** is the MATLAB tool for visual programming and simulation of event-driven systems.

A prominent chapter in the sequel will focus on the use of the MATLAB Symbolic Programming toolbox, which is included in the base product of the MATLAB Student Edition. The Symbolic Programming toolbox of MATLAB is built on top of the *Maple* engine and is a full featured tool for the manipulation and solution of symbolic equations. Treatment in the new book will go up through and include closed form solutions of first-order differential equations.

Finally, curve fitting and an introduction to optimization will be covered in the new book in a problem-based learning solution to a set application problem.

VI   Comments and Suggestions Welcome

Should you have comments, corrections, or suggestions about the current book or about topic coverage in the forthcoming sequel, please contact the editor:

> Great Lakes Press
> Attn.: John Gruender
> PO Box 550
> Wildwood, MO 63040
> E-mail: jg@glpbooks.com
> telephone: 800-837-0201

Also, feel free to visit the Publisher's website at www.glpbooks.com.

Preface to the Second Edition

Although a year can pass by quickly, a lot can happen in 12 short months. Following the introduction of the first edition of **Technical Problem Solving**, our publisher, Great Lakes Press, received a number of helpful suggestions for improvements to the first edition of our textbook. Where ever possible, we have incorporated the suggestions in this second edition. We did not meet all of our targets for the next release of this book (as enumerated in the Preface). We have however quickly turned around three key suggestions leading to three new section in this edition.

First, we have included a problem set at the end of Chapter 2 (A Framework for Technical Problem Solving). Our hope, and our response to a number of requests for this problem set, is that by having concrete problems for which the general problem solving framework can be applied, that we will strengthen in the students' perspective the importance of being systematic in technical problem solving.

Second, in Chapter 6 (2-D and 3-D Plotting and Using MATLAB Help), as the revised chapter name indicates, we have added a completely new section on the tools of three dimensional plotting in MATLAB. This includes both the tools for developing 3-D data plots and the tools from the symbolic toolbox for doing 3-D function plots. Noting that the symbolic toolbox is part of the Student Edition of MATLAB, the ability to do quick and very full featured 3-D function plots leverages MATLAB in way that IS pervasive across engineering curricula.

Third, in Chapter 8 (More Flexibility: Introduction to Conditional and Iterative Programming) we have added a completely new section on **WHILE** loops. In our first edition, we took the pedagogical positions that (a) iterative programming in general is not an easy concept for freshman, (b) for most engineering purposes, **FOR** loops in MATLAB are of more utility that **WHILE** loops, and (c) that spending very substantial time on **FOR** loops (and none on **WHILE** loops) would be of greater advantage to students than spending a little on each. Our user community was almost unanimous in suggesting the inclusion of **WHILE** loops, and we have complied with that strong request.

We look forward to further comments from the user community, and intend to be as flexible as possible in meeting requests for topic additions. Of particular note … in the next edition, which is already under planning, we will add a completely new chapter on symbolic tool box. In addition, your comments will help set our agenda.

Should you have comments, corrections, or suggestions about the current book or about topic coverage in forthcoming editions, please contact the editor:

> Great Lakes Press
> Attn.: John Gruender
> PO Box 550
> Wildwood, MO 63040
> E-mail: jg@glpbooks.com
> telephone: 800-837-0201

Also, feel free to visit the Publisher's website at www.glpbooks.com.

Dr. Jon Sticklen, College of Engineering, Michigan State University
July 10, 2006

Dr. M. Taner Eskil, Isik University, Istanbul, Turkey
July 10, 2006

Chapter 1

Introduction to Technical Problem Solving and MATLAB

Welcome to the world of technical problem solving and to the fabulous problem solving tool MATLAB!

The chapters that follow focus on two general topics: technical problem solving and using MATLAB to help you solve technical problems. This is an introductory book about using MATLAB. Most of the readers have never used a computational environment like MATLAB before, so you have much to master just to become familiar with a core set of MATLAB. But this book is about more than simply how to use MATLAB. As you master MATLAB, you will learn to apply your newfound knowledge to technical problems. In short, this book is about using MATLAB as a tool for technical problem solving.

Think about someone who is learning to be a carpenter. The tools of the carpenter include hammers, screw drivers, saws, etc. When you hire a carpenter to renovate your living room, your assumption is that the person you employ will know how to use the various tools and when to use each. No matter how good the screwdriver skills of the carpenter, if the task entails cutting a 12-foot 2x4 into two 6-foot pieces, you would be dismayed if your carpenter attempted to make the necessary cut by repeatedly punching his screwdriver into the 12-foot length of wood. Of course, he should use a saw for that task. The analogy may be a little strained, but many students learn how to use MATLAB at the nuts and bolts level yet have difficulty in choosing the appropriate tool within the MATLAB environment to apply to a specific problem.

When you complete the material in this textbook, you will be able to utilize a core set of MATLAB tools and know which tools in that core set to use for a given problem.

1-1   What is Technical Problem Solving?

Technical problem solving is good common sense applied to technical problems, which leads to the question "What is a technical problem?" A clue lies in the places where we find problems that most of us would term *technical*. Areas like the natural sciences (physics, chemistry, biology, etc.), engineering (mechanical engineering, civil engineering, chemical engineering, electrical engineering, etc.), and some areas of the business professions (accounting, finance, etc.) typically entail problems that most people would call *technical*. The question is "What are the general characteristics of problems from these disciplines?"

Problems from such areas are quantitative in nature. Quantitative problems are problems which demand a numerical solution or which require numerical computations to reach a solution. A problem like "How many pounds of liquid oxygen will be required to propel the Space Shuttle to orbit?" requires you to produce a number as a result. Likewise, when a structural engineer is asked "How many cars will the new Tacoma Narrows Bridge safely carry?" an acceptable answer is not "Quite a few." Though not strictly true, for purposes in this book, we equate technical problem solving with quantitative problem solving.[1]

We need to pursue this quantitative face of technical problem solving a little further. Especially in engineering, quantitative problem solving provides the basis for making many decisions. Consider the following question posed to NASA officials every time a Space Shuttle is scheduled for launch: "Should we go for launch today?" The direct answer is either Yes or No, and neither answer is a number. But many (perhaps thousands of) numerical computations have to be completed before the NASA engineers come to a Yes/No answer for a particular Shuttle launch. For any Yes/No decision in a technical area to be believable, it must be justifiable. That is not to say that human judgments based on human values do not enter the equation; they most certainly do.

In addition to being rooted in numerical calculations, technical problem solving is usually complicated enough so that the complete problem must be broken down into a number of steps. If I told you that eggs cost $1.00 a dozen, and asked you how much would five dozen eggs cost, your result would be a numerical answer. But the problem is so simple, there is no need to break down, or *decompose*, the problem. On the other

---

1. One problem-solving approach that typically does not focus on quantitative results is artificial intelligence (AI). Knowledge-based systems or expert system, are computer programs that typically offer qualitative support for decision making, including decision making in technical areas.

hand, if I ask you to redesign the Super Turbo 348 Porsche engine so that the revised 348 will attain 15 percent higher top speed, then more than likely you will not produce an answer without going through many systematic steps. Part of how well you do your job of redesigning that Porsche engine lies with how well you decompose the problem.

Neither of our criteria used to pin down technical problem solving (numerical orientation and reliance upon problem decomposition) are hard and fast. Technical problem solving is not so easily defined. At first blush, it is one of those things that "you just know when you see it." For example, analyzing a newly developed polymer composite for possible structural use in a bridge would definitely fall into the realm of technical problem solving. Likewise developing a simulation model of stellar evolution based on known physical laws with the goal of making predictions about a star's life history as a function of its mass would be considered technical in nature. Accomplishing either of those tasks would be an easily accepted example of technical problem solving.

To end this discussion of the definition of technical problem solving and to provide you a working target to use as background for your studies using this textbook, you can think of technical problem solving as the mainstay of what an engineer does and what you must learn to do if you are to become an engineer.

1-2   The Path to Becoming a Good Technical Problem Solver

If becoming a good technical problem solver is part and parcel of becoming an engineer, what must you do to become a good technical problem solver? There are two general elements. First, you must master the subject matter of a given technical area to become a good problem solver in that area. If you plan on becoming a mechanical engineer, you must master those conceptual topics that are assumed knowledge for any beginning mechanical engineer. Similarly, if you are planning on becoming a chemical engineer, you must master certain topics in your field of interest. No matter what direction you are headed, you must master the conceptual subject matter.

The question then becomes how do you demonstrate that you have learned the conceptual topics in a given major? In technical areas, the answer is almost always that you show you can solve problems related to that major. For example, no matter how well civil engineers can describe an earthen dam, they will not have a great professional future if they cannot compute the thickness to which a given earthen dam should be built. *The proof of the pudding is in the problem solving.*

And that bring us to the second general element you need to be a good technical problem solver. You must master the tools of the trade. In former times, a tool that every engineer was expected to use competently was the slide rule.[2] These days, the nature of tools for working engineers has moved to computational, computer environments. Examples include MATLAB, Mathematica, and MathCad, to name a few. All of these computational environments include support for setting up and performing numerical computations.

The two ingredients for becoming a successful engineer—mastering the conceptual subject matter of your chosen major and mastering the tools that will help you solve problems in your major area—go hand in hand.

---

2. In Google, type in "slide rule" and click on the link labeled "Java Slide" to see what a slide rule looks like.

1-3   Your Part of the Bargin

The section above was written in general terms. This section focuses on the specifics of what you, the beginning learner, must do to master this subject matter. If you follow the prescription below, you will have a working knowledge of the rudiments of MAT-LAB and an expanded sense of technical problem solving.

First, plan on spending substantial time learning this material. If you are using this book in conjunction with a formal class, as a rule of thumb plan on spending at least two hours studying outside of class for every hour you spend in class and lab. But ask your instructor what is recommended for your particular class.

Second, from Chapter 3 on you will need to read assigned sections of this text with MATLAB at your side, since those chapters specifically cover MATLAB topics. Experience has shown that students who do not write out the examples in the reading in MATLAB (who do not actually try to use the operations/code of the examples) tend to view learning MATLAB as a big set of memorization tasks. MATLAB cannot be learned by that method. If you engage yourself as you read by working with MATLAB as you read about it, the learning will come naturally.

Third, work the problems that your instructor recommends or assigns to you at the end of each chapter. You cannot learn MATLAB effectively if you do not work a substantial number of problems using MATLAB! Problems are organized into two groups at the end of each chapter beginning with Chapter 3. Set A problems teach you how to use MATLAB. Set B problems teach you how to apply MATLAB to specific situations.

Fourth, after you complete each chapter, step back and reflect on your own learning. In particular, as part of that reflection, ask yourself these questions:

- What was this chapter about? (Answer this question in a short narrative paragraph. If you find yourself listing the headings or listing the individual MATLAB commands of the chapter, then you are missing the point. The idea is to develop a framework for yourself in which MATLAB makes sense to you.)

- How does this chapter extend what I knew from previous chapters? (Again, frame your answer in a short paragraph. Pay particular attention—from Chapter 3 on—to the widening sense of what you can do with MATLAB.)

If you put in the time to read assigned sections and to work assigned problems, you will find that your effort will pay substantial dividends. Follow the four points of the prescription above, and you will be well on your way to masterings MATLAB.

Chapter 2

A Framework for Technical Problem Solving

This chapter describes a relatively complete framework for technical problem solving. Almost every textbook in freshman engineering starts with such a framework. Remember as you read this chapter that this, and every other general technical problem-solving framework, is a guide rather than a set of laws. Nothing is etched in granite here.

Remember also that the more complex a problem is, the more systematic you need to be in working toward a solution. In a nutshell, this is why any framework for problem solving is useful: The framework provides a guide for the steps that may be useful in working from the initial problem statement toward a solution. You will not need every one of the steps below for every technical problem you solve. But you will use the steps below as a checklist for steps that may be useful.

In this chapter, we will introduce the steps of a framework for technical problem solving and briefly explain each. Then we will go on to a detailed example that uses the framework.

2-1   Steps in a Framework for Technical Problem Solving

Step 1: Refining the problem statement and structuring the problem

Often when you are first given a problem statement you will find that what you are being asked to solve is not clear. This is not because your instructor (boss, client, etc.) is too lazy to describe what they want, but because, in a realistic and reasonably sized project, the goal is rarely defined precisely. You will need to refine the problem statement to arrive at a precise problem statement, one with which you can move toward a solution. And you will need to give the problem initial structure by stating what the problem inputs are and what computational output is expected. This, of course, does not mean that you are supposed to give the explicit numerical output as part of the first step. It does mean that you should list the input and output variables of the problem.

Step 2: Making a sketch or diagram that helps you understand the problem

One of the most common traits of seasoned technical problem solvers is that they can visualize the physical situations for problems. Having a visual image that represents a physical situation can be a tremendous aid in sorting out the relationship of variables in a problem and can aid in understanding the entire problem from a high-level perspective. As a stepping stone on your path to having these visualization abilities, you must get in the habit of making a sketch or diagram of a problem as you begin to solve it.

Step 3: Assembling and organizing background knowledge

After you have refined your problem statement and you have visualized the problem, the next question you need to ask yourself is what you need to know to solve the problem. In reality, this can be a difficult step. How many times have you said, "If only I had known..." about some fact that could have turned a hard problem into an easy one. Once you have determined what you need to solve a problem, then you need to get the required information. The Internet and its associated search engines, like Google or Yahoo, have made that task easier. Finding material that would have taken days to locate in a library can now be found on the Internet rapidly.

Step 4: Making appropriate simplifying assumptions and approximations

Suppose you are asked to find the acceleration that a prototype of a new motor scooter can achieve. You are told that this will be a low cost vehicle with a top speed of only 25 miles per hour. Given all you know, you search out the appropriate equations and apply them. If you include the effect of air resistance in your computations, you will have made your problem harder than it need be. Air resistance does not play a significant role in this situation because of the low speeds involved. Finding an appropriate approximation can simplify problem solving. You must make assumptions or approximations you use in technical problem solving explicit. If you do not, then others may apply your results in situations where the assumptions/approximations do not hold.

Step 5: Decomposing the problem to a series of simpler problems

Think about how you understand an automobile. Is it one large machine with thousands of individual parts? In reality, an automobile is exactly that: a complex machine with many working parts. When automotive designers are asked to develop a new model, they do not start by thinking about every single part of the car but rather focus on the major systems of the car: drive train, fuel system, cooling system, etc. By decomposing the design problem into a number of smaller design problems, the complexity of the problem solving is reduced. For complicated technical problems, decomposition of the problem into smaller problems is essential.

Step 6: Performing a dimensional analysis

Dimensional analysis is a quick and easy method for determining whether the mathematical relationships you intend to apply in your solution are flawed. In dimensional analysis, you substitute for each variable of a relationship the units of the variable, then algebraically simplify the relationship. If the relationship you intend to apply is accurate, the simplification should yield an identity of the units.

Steps 1-6 are aimed at taking an initial problem statement and turning it into a problem (or series of sub-problems) that can be readily solved.

Step 7: Computing the detailed answer and discussing the results

Once a technical problem has been sharpened and readied for solution, then a computational tool like MATLAB may be used to perform the computations needed to obtain the solution. The final step then is to examine the results, understand them, and be ready to explain those results. This last step is critical: Without examining and understanding the results generated using MATLAB (or any other computational tool), your job is not done. Understanding your results is perhaps the most important step in the entire framework of problem solving.

In seven steps you have a framework for developing a solution to a technical problem. At this point, you may feel somewhat uncomfortable because though these abstractly stated steps may sound fine, you may feel quite unsure how to apply them at this point. That is the topic for the next section in which the framework will be applied systematically to a common physics problem: determining projectile trajectories. By the time you finish the next section, you should understand the framework above and the need for some anchoring framework to act as a guide for technical problem solving.

2-2   An Example Using the Framework

In this section we will methodically apply the framework for technical problem solving that we sketched. The problem we are going to solve using our framework is a straightforward application of beginning physics.

> **What is the optimum firing angle we should set for a catapult whose purpose is to hurl hay bales to a herd of starving caribou, given the initial velocity of hay bales as they exit the catapult?** [1]

2-2.1   Refining and Structuring the Problem

The first step in our framework for technical problem solving is *structuring the problem and refining the problem statement*. As we discussed, structuring the problem means identifying the input for the computation that leads to a solution, and identifying the output from the computation.

Even though the problem statement for the hay bale problem is straightforward, it still has a glaring ambiguity in it and needs refinement: What is the meaning of "optimal"? Does the problem statement mean we are to find the launch angle that will maximize the height the hay bale will reach? Or does it mean we are to find the launch angle that will maximize the horizontal flight distance of the hay bale? Or does it mean something altogether different? This illustrates the importance of understanding the beginning problem statement. To clarify a problem statement, the ultimate source of information is the person setting the problem, the client in the typical case of an engineering problem. A second source is the intelligence between your ears.

In our hay bale problem, common sense understanding of the problem is enough to clarify the problem statement. The maximum vertical height a hay bale reaches is of no real consequence in the situation while the maximum horizontal distance to which we can hurl the hay bale is relevant. Hence, let us refine the problem to this:

---

1. If you do not know what a catapult is, then take it as a mechanical device designed to throw some material object. Catapults are interesting mechanical devices among a family of physical devices honed during the Middle Ages. Predating the rise of modern physics, the development and improvement of such devices provided a great deal of experience with mechanical devices. With experience gained in the construction of the great cathedrals of Europe during the same broad time period, this knowledge set the stage in part for the emergence of generalization and explanations offered by Newtonian physics. If you are interested in finding more on this subject, use a good web search engine and search for "catapult."

What is the catapult's firing angle needed to hurl projectiles a maximum horizontal distance given the initial velocity of the object?

In addition to clarifying the initial statement, we have also generalized from "hay bale" to "projectile." No "Newton's law of hay bale motion" exists but a set of "laws" exist in physics that can be used to describe the motion of all objects in a Newtonian world. We can make this generalization of the problem statement because it makes no difference in terms of the solution methods whether we are hurling hay bales or snowballs.[2] Both are covered under the general term *projectile* and more importantly, our background knowledge of how objects move in the world is indexed by generalized terms like projectiles. Test this by taking out your elementary physics textbook and looking up the entry "hay bale" in its index. Unless you have an odd physics text, you will not find an index entry for "hay bale." Now look for the word "projectile" in your physics text's index, and you will probably find it. There is a very important lesson here.

We index our background knowledge of the world by terms that are general. Why? Consider the hay bale problem. An index that included the name of every specific object that we could throw (by whatever means) would be outrageously long (e.g., hay bale, flaming ball of wax, rock). Instead we capture the general sense of *object thrown* by the single term *projectile*. All of this is in the abstract. What is the concrete impact on *you*? It is that *you* must develop an ability to generalize from a particular problem situation (like "hay bale thrown from a catapult") to the general version of the problem ("projectile thrown"). Once done, you can find the general background knowledge that applies to your situation. If you cannot recognize a given specific problem as an example of some general problem statement, it is likely you will not be able to solve the problem.

At this point, we can move on to nailing down the input(s) and outputs(s) of the hay bale problem. The input is easy to identify. The only variable we can set is the launch angle of the projectile since the initial speed of the projectile is fixed. We need to make the input and output variables explicit, that is, we need to name them because it is hard to talk about something you do not have a name for. Let us assign the variable ϕ to the trajectory angle (in degrees). The output variable of the problem is similarly easy with one output variable: the total horizontal distance the object will travel. Let us assign the variable D to be the total horizontal distance the object travels (in feet).

---

2. It makes *almost* no difference. We will deal with this again when we discuss assumptions for this problem.

One input variable ϕ (in degrees) and one output variable D (in feet) — Right?

Wrong! (Give yourself a pat in the back if you caught it.) Read the problem statement again. We are supposed to find the trajectory angle that will maximize the horizontal flight distance of the object. The output of the computation is an angle. The lesson is to learn to read the problem statement and its refinements carefully and be sure that your statement of the Problem Solving Task (the input-output relationship) of the problem matches the problem statement.

The input variable to the computation solving the problem is the value of the initial velocity v_0 of the projectile; the output variable from the computation solving the problem is the angle $\phi_{\text{maximizing}}$ that will maximize the horizontal travel distance D of the object. Putting this into mathematical language, our problem is to find the computational procedure or, in alternate language, the function g that computationally maps from v_0 to $\phi_{\text{maximizing}}$:

$$\phi_{\text{maximizing}} = g(v_0)$$ (EQ 2-1)

2-2.2   A Sketch or Diagram of the Example Problem

Figure 2-1 is a hand sketch of our problem. A sketch of a problem situation is incomplete without a short narrative explaining the sketch. L_1 is a vertical line drawn upward from the launch point for the projectile, L_2 is the launching direction of the projectile, v_0 is the initial velocity of the projectile, and h is the maximum height the object reaches during its flight (measured in feet). The flight trajectory is the heavy curved line in the sketch, D is horizontal distance the object travels (measured in feet), and ϕ is the launch angle in degrees measured from the horizontal. The small circle on the trajectory is our projectile in flight at our instant in time.

Why spend all this time on making and explaining a sketch for such a simple problem? Everything was stated in the final version of the problem statement, right?

Wrong! Congratulations if you caught it and answered No.

Go back and read the last version of the problem statement. The problem statement reads in part *What is the firing angle?* Look at the sketch in Figure 2-1. The angle we have used for the firing angle is the angle marked ϕ on the diagram. But another choice could have been the angle indicated from the vertical axis to the firing direction. By completing a sketch of the problem, we helped to make our

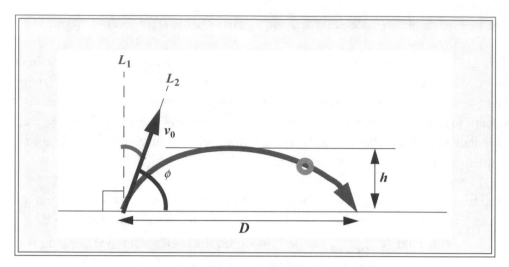

Figure 2-1: Sketch of the Hay Bale Problem

understanding of the problem concrete, and solidified our use of variables. By making the sketch and concisely writing a description of it, we straightened out a hidden ambiguity we did not even notice before: the possibility of representing the firing angle either by ϕ or its complement. This is an example of one stage of technical problem solving (making the sketch) feeding back to a prior step of problem solving (structuring and refining the problem). We will see this characteristic of standard technical problem solving again: feedback and interaction between the various steps. Such interaction between the steps of problem solving is a general trait of technical problem solving.

Realizing now the ambiguity that was in the earlier version of the problem statement, we need to revise what angle we are using.

> **What is the firing angle measured from the horizontal we should set for a catapult to hurl projectiles a maximum horizontal distance given the initial velocity of the object?**

Making a sketch of the problem situation is an example of visual reasoning, i.e., visualizing a problem situation and reason on the basis of it. Visual reasoning is an indispensable part of technical problem solving. Visualization can be a mental process with no physical sketch drawn. For seasoned engineers the step of making a sketch is often skipped. A seasoned engineer often thinks in terms of mental sketches. For novices in technical problem solving, physically constructing a sketch in detail sufficient to show all aspects of the problem is necessary. As an added benefit, having a sketch of your problem situation makes communicating your understanding of the problem easier, whether to a colleague, your boss, or your academic class instructor.

2-2.3   Background Knowledge for the Example Problem

Like most technical problems, there is more than one solution path for our hay bale problem. Here, we work toward a solution using (a) general knowledge of physics and (b) general knowledge of one way in which optimizing problems may be solved. Our hay bale problem is an optimization problem. We seek the value of the firing angle ϕ that maximizes the total horizontal distance D traveled given an initial speed of the hay bale as it comes out of the catapult.

The other general type of optimization problem is one in which we seek to minimize a problem variable; we will encounter a number of optimization problems of maximizing and minimizing types over the course of this book. Optimization problems are often encountered in technical problem solving. For example, the design of aircraft can boil down to minimizing the weight of the aircraft while simultaneously meeting constraints for required aircraft thrust and lift. Consider the problem of an automotive assembly plant meeting production quotas while minimizing the inventory of parts used. The list of examples is long. One way optimization problems can be solved involves a three-step process that is spelled out in Figure 2-2 for our hay bale problem.

---

1. **First we note that the variable we want to optimize is a function of one of the input variables in the problem. In the hay bale problem the total horizontal distance is the optimized variable and the initial launch angle is the optimizing input variable: $D = f(\phi)$.**

2. **Second, we sample values of the optimizing input variable and for each compute a value for the optimized variable. That is, we sample from the possible values of ϕ, and compute corresponding values for D. Suppose that the sampling we take over ϕ is the sequence $\phi_1, \phi_2, \phi_3, \ldots \phi_N$. For each of these angles, there is a corresponding value of the horizontal distance traveled $D_1, D_2, D_3, \ldots, D_N$ that can be computed using the function $f(\phi)$.**

3. **Third, we examine the results of applying $f(\phi)$ to the angles in the sample, that is, we examine the values of $D_1, D_2, D_3, \ldots, D_N$. We choose the value of the optimizing input variable ϕ that maximizes the value of the optimized variable D.**

Figure 2-2: Procedural Steps for Solving the "Hay Bale" Optimization Problem (function g in Equation 2-1)

There are two complications here. We need to choose a good sample of the optimizing input variable. We will discuss this issue in a later chapter. For now, see if you can understand what good sample means in this context. Hint: If I choose a sampling of ϕ to be 1, 2, 3, ..., 10 degrees, I would not have a good sample. A second complication is the (likely) situation that the value for ϕ that produces the maximum D is not in my explicit sample of possible ϕ values. This second complication involves a strategy (and a technique to implement it) called interpolation.

You must keep two functional mappings straight to understand this solution method. The first task is to develop the function (or computational procedure) g specified in Equation 2-1. The computational procedure for g is in Figure 2-2. So, we have the first task completed.

The second task is to develop the function f introduced in Step 1 of Figure 2-2. This is where general background knowledge of physics enters the picture. The physics background knowledge we need for our problem boils down to five items.

1. We need to know that we can analyze the horizontal flight of the projectile independent of the vertical flight of the projectile.

2. We need to know that the initial velocity v_0 of the projectile in a direction that is ϕ degrees from the horizontal can be thought of as composed of two components: one in the horizontal direction $v_{0,horizontal}$ and one in the vertical direction $v_{0,vertical}$. Simple trigonometry can give us the values for $v_{0,horizontal}$ and $v_{0,vertical}$ given the value of v_0 and ϕ.

3. We need to know only one force is on the projectile once it leaves the catapult, and that force is acting in the vertical direction to pull the projectile toward the ground. Further, this gravitational force produces an acceleration of 32 ft/sec^2 in the vertical direction toward the earth. We will give this constant the name g. A version of Figure 2-1 which includes the gravitational acceleration is shown in Figure 2-3.

4. We need to know that in general, given an initial velocity v_0, and a constant acceleration a, the velocity $v(t)$ of a projectile at a time t is given by the following:

$$v(t) = v_0 + at$$ (EQ 2-2)

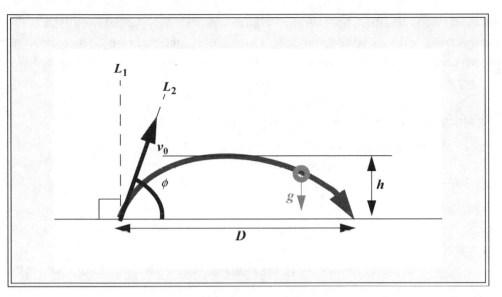

Figure 2-3: Augmented Sketch of the "Hay Bale Problem" Including the Acceleration on the Projectile

5. We need to know that if there is no acceleration, then the distance d that a projectile will travel if it is going at a velocity v over a time t is given by

$$4d(t) = vt$$ **(EQ 2-3)**

We now have all the pieces of technical background knowledge.

Let us be clear where we are:

1. We have a computational procedure for solving the maximizing problem we are working on. This computational procedure is specified in Figure 2-2.

2. We have assembled the background knowledge of physics, indexed by the generalization of our hay bale problem to a projectile problem.

3. We know that working forward from this point, we need to develop a function $D = f(\phi)$ that we can apply to a sample of ϕ values to produce a series of D values from which we can find the ϕ that maximizes D.

We could alter #3 and, instead of working from general physics knowledge: look for (or remember) the explicit physics relationship that will give us $\phi_{maximizing}$. However, it is common in technical problem solving not to find the relationship that gives us what we need. We will utilize the background knowledge above by algebraically manipulating what we have to produce what we need: the function f. Once we have f, the description of our solution method in Figure 2-2 on page 2-15-2, will be complete, and we will be ready to apply it to the hay bale problem.

The algebraic manipulations are shown below in step-by-step form:

1. Write down the version of Equation 2-2 that is correct for the vertical motion of the projectile:

$$v_{\text{vertical}}(t) = v_{0,\text{vertical}} - gt \qquad \text{(EQ 2-4)}$$

   There is a minus sign for the gravitational acceleration term because the gravitational acceleration is in the downward direction.

2. Write down the component of the initial velocity v_0 that is in the vertical direction:

$$v_{0,\text{vertical}} = v_0 \sin(\phi) \qquad \text{(EQ 2-5)}$$

   We know the initial velocity of the projectile v_0 because it is a given. We know the value of ϕ because it will be one of the sampling set we will set up. Hence, we know the value of $v_{0,\text{vertical}}$.

3. Analyze the vertical motion of the projectile to find how long it takes to reach its highest point: h in Figure 2-3. Apply our intuitions of the physical world and understand that the velocity of the projectile in the vertical direction will be zero at the highest point the projectile reaches. Since the velocity of the projectile is zero at the high point in the trajectory, it follows that to find the time to reach the highest point we can set the left-hand side of Equation 2-4 to 0. Further, we can substitute the value of $v_{0,\text{vertical}}$ from Equation 2-5 into Equation 2-4. The two operations give us the following at the time t_{highest} at which the highest point is reached by the projectile:

$$0 = v_0 \sin(\phi) - gt_{\text{highest}} \qquad \text{(EQ 2-6)}$$

   Now, we solve for t_{highest}:

$$t_{highest} = \frac{v_0 \sin(\phi)}{g} \qquad \text{(EQ 2-7)}$$

4. Again relying on physical intuition if the time to reach the highest point in the trajectory is $t_{highest}$, then the **total** flight time t_{total} will be twice as much. (It takes an equal time to go up as it does to come down.) Thus, the total flight time of the projectile will be the following:

$$t_{total} = 2\left(\frac{v_0 \sin(\phi)}{g}\right) \qquad \text{(EQ 2-8)}$$

5. Applying simple trigonometry, the velocity in the horizontal direction is the following.

$$v_{0,\,horizontal} = v_0 \cos(\phi) \qquad \text{(EQ 2-9)}$$

We can apply Equation 2-3 and find the horizontal distance covered in the entire flight of our projectile:

$$D = \frac{2v_0^2 \sin(\phi)\cos(\phi)}{g} \qquad \text{(EQ 2-10)}$$

We have what we set out to get, which is the total horizontal distance covered by our projectile as a function of the firing angle ϕ. The other terms in Equation 2-10 (the initial velocity of the projectile and the acceleration due to gravity) are constants.

6. One more step is to choose a sample value set of ϕ, apply Equation 2-10 to each value in the sample value set, and determine what value of ϕ will maximize the value of **D**.

We will leave the operations in Step 6 for Section 2-2.7 on page 2-24.

2-2.4 The Assumptions and Approximations of the Example Problem

In the preceding sections, we developed a series of steps to solve the hay bale problem. Following our general framework for technical problem solving, we need more steps before we are ready to implement the solution. Assumptions and approximations used in our solution for the hay bale problem are covered in this section. In Section 2-2.6, we will discuss the dimensional analysis appropriate for the problem.

The line between an assumption and an approximation is hard to determine. In many cases, the difference is more syntactic than semantic, i.e., more an issue of how the words you use describe the assumption/approximation. You need to keep this in mind as we work through this section.

Let us start by referring to Footnote 2 on page 2-12. In the discussion about generalization of problem terms to enable finding the appropriate physical principles to apply to a problem, we said that it made no difference in our hay bale problem whether the catapult was hurling hale bales or snowballs. But, at a precise level of measurement, it would make a difference. The physical principles we applied were derivatives of Newton's First Law and Second Law. Newton's First Law asserts that objects (including projectiles) will continue in their current state of motion if there is no force acting on the objects. The horizontal motion of the hay bale is governed by Newton's First Law because there is no force acting on the hay bale in the horizontal direction. Newton's Second Law asserts that if a force is acting on an object, then the acceleration the object experiences is inversely proportional to the mass of the object ($a = F/m$). The vertical motion of the hay bale is governed by Newton's Second Law because the force of gravity acts on the hay bale in the vertical direction.

Let us examine the application of Newton's First Law to the horizontal motion of the hay bale. If there is no other influence on the object, the object will remain in its current state of motion. Our application of Newton's First Law to the hay bale problem assumes there is no influence on the hay bale that would alter its current state of motion. Is the assumption 100 percent valid? No! One way to get a handle on the validity of assumptions is to push the situation of the problem to an extreme. For the hay bale problem, one way to push the problem to an extreme is to imagine varying the initial velocity of the hay bale out of the catapult to higher values. Suppose we pushed the initial velocity of the hay bale to 2,000 feet/second. Clearly pushing to this extreme would require a physically strong catapult. But that is not an issue: This is a mental exercise meant to test the validity of the assumptions of our solution method.

When you push a problem to its extremes in this way, you have to apply your own knowledge to determine the outcome of the extreme situation. So, is there anything from your own experience or knowledge that tells you what happens as objects move rapidly through the air? Consider the fact that NASA Space Shuttles, on reentry to the Earth's atmosphere, become extremely hot until velocity of the craft decreases. The reason is because of the resistance of the air as the Shuttle moves at speeds greater than the speed of sound. The air causes frictional forces on the craft that heat its outer shell: the higher the velocity, the greater the frictional forces. In our everyday lives, we rarely consider air resistance because velocities of objects in everyday life typically are not great enough to make this effect important.

If our hay bale problem included hurling the hay bale at velocities above, say, 75 feet/second, then the air resistance assumption that underlies our solution would not be met. In that case, we would need to consider the effects of air resistance on the trajectory of the hay bale. How about at velocities below 75 feet/second? The assumption is still not met perfectly. Assuming there is no appreciable effect due to air resistance at velocities below 75 feet/second is a reasonable approximation. This discussion is an example of the dualism of approximations and assumptions: We assumed negligible effects due to the resistance of the air to develop our solution; the assumption is not perfectly valid, but it is a good approximation.

By making the assumption of negligible air resistance, we simplified the problem and enabled an easy solution. There is a strong lesson here for those who do technical problem solving. If we had started out with the harder version of the problem including air resistance, it would have made the physical situation more complex and the solution much harder to achieve. Many times in technical problem solving, the path to a solution leads to making assumptions about the problem and solving a simplified version of the problem. Then the solution for the simplified case can be used as a starting point to develop a more precise solution. For example, suppose the person giving the hay bale problem to us looked over our solution including the assumptions under which it is valid (velocity of the hay bale out of the catapult is less than 75 feet/second) and then informed us that he or she intended to employ a catapult whose release speed was 200 feet/second. We would have to rework the solution, though not from scratch. All of the solution path we followed would still be partially correct. What we would need to do is to step back through that solution path and augment each part of it to include the effect of air resistance. "Augment" implies we build on our first solution.

Often, technical problem solving takes such an incremental approach in which we make simplifying assumptions to enable a quick and dirty solution to a problem as a first step. Even if these assumptions are not perfectly accurate, the solution developed may be a good enough approximation for the intended purpose to allow it to stand, such as for catapult velocities less than 75 feet/second. If the problem setting we are interested in violates the assumptions used to such a degree that the solution produced using the assumptions is inaccurate by an intolerable amount, then incremental improvements to the initial solution are made by removing assumptions one by one and reworking.

The most important aspect in this section is that you must be aware of the assumptions you are making in a problem solution, and you must communicate the assumptions for your solution to the person who originally set the problem for you. The problem setter who is most close to the problem being solved is most likely to know if the assumptions made are realistic for the problem and, hence, likely to lead to a solution that will be accurate.

As an additional point for this section, a characteristic of the height that the projectile reaches involves another assumption embedded in our solution for the hay bale problem. See if you can find the part of our solution that depends on an assumption of projectile height, and see if you can determine how you would incrementally change our solution to relax that assumption.

Finally, to end this section, let us again consider Footnote 2 on page 2-12 in which we indicated that it should make a difference whether we are hurling hay bales or snowballs. How would you reason toward answering the question "Which will go farther when hurled at the same launch angle and initial velocity from a catapult: a snowball or a hay bale?" Hint: Start with the solution we have developed for no air resistance, and, in the abstract, figure out if the projectile will go further with no air resistance or with air resistance included. Then develop an argument about whether a snowball or a hay bale would suffer more air resistance during its flight from the catapult.

2-2.5   Decomposing the Problem / Recursive Structuring of the Example Problem

The idea of recursive structuring of a technical problem is to break the problem into pieces, each with its own well-defined input and output, and go on to develop technical solutions for each piece.

In Section 2-2.1, after a false start, we identified the single input variable for the hay bale problem as the initial velocity v_0 of the hay bale as it leaves the catapult. Likewise, we identified the single output variable as the initial angle from the horizontal $\phi_{\text{maximizing}}$ that maximizes the horizontal flight distance D (for a given value of v_0). The computational procedure we gave for such optimization problems is in Figure 2-2. That solution, however, involved finding another computational relation f that would take the initial trajectory angle from the horizontal ϕ as a single input and produce the value for the total horizontal flight distance D as a single output (for a given value of ϕ).

The above description is of a recursive problem solving procedure. We developed (or looked up) a solution for optimization problems (Figure 2-2). That was the higher-level solution to the hay bale problem. But in that solution, we produced another problem since the solution involved the function f. Hence, to solve the hay bale problem, we have to provide a solution for f.

Though our solution for the hay bale problem does involve a recursive structuring, it is relatively simple and straightforward. The true power of the idea of problem decomposition and recursive problem structuring is most apparent for more complex problem as will be seen later in this book.

2-2.6   Dimensional Analysis for the Example Problem

The solution methods we have developed for the hay bale problem are the following:

- The computational procedure in Figure 2-2
- The relation shown in Equation 2-10

Dimensional analysis is useful as a check on the validity of Equation 2-10.

The result of the algebraic manipulations of the background physics knowledge yielded the following:

$$D = \frac{2 v_0^2 \sin(\phi)\cos(\phi)}{g} \qquad \textbf{(EQ 2-11)}$$

This is a restatement of Equation 2-10. Here are the units (dimensions) in this relation:

- D is a distance in units of **ft** (feet)
- g is the constant acceleration due to gravity in units of **ft/s^2** (feet per second^2)
- v_0 is the initial velocity in units of **ft/sec** (feet per second)
- ϕ is the initial trajectory angle measured from the horizontal in units of **degrees**

Using these units, noting that the sine and cosine function produce pure numbers (no units), and applying dimensional analysis, we obtain Equation 2-12. Dimensional analysis starts with an algebraic equation that we think is correct, filling in the units for each of the equation's variables, and seeing if we end up with an algebraic identity:

$$ft = \frac{\left(\frac{ft}{s}\right)^2}{\frac{ft}{s^2}} \qquad\qquad \text{(EQ 2-12)}$$

$$ft = \frac{ft^2}{s^2} \times \frac{s^2}{ft} \qquad\qquad \text{(EQ 2-13)}$$

Either of the preceding two equations yields the following:

$$ft = ft \qquad\qquad \text{(EQ 2-14)}$$

Thus, dimensional analysis shows that Equation 2-11 is consistent for units. Dimensional analysis can be easy. In more difficult problems, dimensional analysis remains easy to apply, and the insights it offers and the errors it uncovers can be important.

2-2.7   Putting It All Together: The Solution for the Example Problem and Discussion of Results

Finally, we are ready to bring together a solution for the hay bale problem. We follow the computational procedure shown in Figure 2-2. We need a sample set of values for ϕ. From that sample set, we compute the value of D using Equation 2-11. For values of ϕ that range over [0, 10, 20, ..., 90] degrees, Table 2-1 gives the results for D, for an initial velocity of 50 feet/second and gravitational acceleration at sea level is 32 feet per $second^2$.

We can do better than this by generalizing the table; we can make a similar table that will be accurate for any value of initial velocity. The initial velocity v_0 appears in Equation 2-11 as a simple multiplying factor. Given one value of v_0 there is one value of D; for v_0 of 50 feet per second, the values are shown in Table 2-1. Given any value of v_0, Table 2-2 lists generalized results.

The function that computes D from ϕ behaves well, that is, the data in Tables 2-1 and 2-2 increase smoothly up to between 40 and 50 degrees, then fall regularly down to reaching 0 feet again at 90 degrees. A good tactic at this point is to plot the results in

Table 2-1: Values for Total Horizontal Distance (initial velocity = 50 ft/sec)

ϕ (degrees)	D (feet)
0	0.0000
10	26.7203
20	50.2178
30	67.6582
40	76.9381
50	76.9381
60	67.6582
70	50.2178
80	26.7203
90	0.0000

Table 2-2: Values for Horizontal Distance (initial velocity = 1 ft/sec)
(multiply by initial velocity squared to obtain total horizontal distance)

ϕ (degrees)	D (feet)
0	0.0000
10	0.0107
20	0.0201
30	0.0271
40	0.0308
50	0.0308
60	0.0271
70	0.0201
80	0.0107
90	0.0000

Table 2-2.[3] The immediate goal is to find the value of ϕ that maximizes **D**. From the table in Figure 2-2 and the graph in Figure 2-4, the maximizing value of ϕ is between 40 degrees and 50 degrees. From the symmetry of the curve in Figure 2-4, it is appropriate to conclude that the $\phi_{\text{maximizing}}$ is 45 degrees.

Here, we restate our final version of the problem statement:

---

3. We will examine graphing in MATLAB. The typical goal of the work-to-come focus on making plots is best described as plotting in the service of problem solving. That is the sense we use the plot in Figure 2-4.

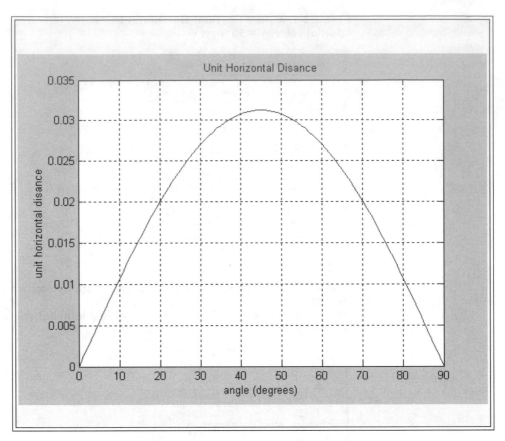

Figure 2-4: Graph of Data from Table 2-2

What is the firing angle measured from the horizontal we should set for a catapult to hurl projectiles a maximum horizontal distance given the initial velocity of the object?

We have now completed a solution and found an answer for this problem. The initial angle of firing should be set to 45 degrees from the horizontal to obtain the maximum horizontal distance for the projectile. We have found that this maximizing firing angle is independent of the initial velocity imparted to the projectile. We have gone further than the problem statement asked of us because the plot in Figure 2-4 allows us to conclude that the maximum distance achieved by any projectile fired at sea level, at an initial angle of 45 degrees, and with an initial velocity of v_0 will be the following:

$$D_{maximum} = 0.031 v_0^2 \qquad \text{(EQ 2-15)}$$

The solution in Equation 2-15 is subject to the assumptions we have made. The one assumption/approximation we discussed in Section 2-2.4 was a limitation on the launching velocity of the projectile due to effects of air resistance. Other assumptions (as already hinted at) include an assumption about being at sea level.

Here is a list of points you should take away from the discussions of this section:

---

1. In technical problem solving, you must identify input and output variables correctly from the beginning problem statement.

2. Creating a sketch of a problem situation is necessary in technical problem solving. The utility of sketch making lies in crystallizing understanding of a technical problem by visualizing it.

3. A good technical problem solver has the ability to generalize a problem situation to obtain problem terms that index general knowledge to be applied to the problem.

4. A good technical problem solver has an ability to manipulate general background knowledge to obtain relationships between physical variables that apply directly to the problem situation.

5. A general approach to optimization problems is that shown in Figure 2-2.

Synopsis for Section 2-2

2-3   Problem Set for this Chapter

The problem sets that are at the end of each of the later chapters in this book are divided into two sections. Set A problems will focus on "nuts and bolts" issues of the chapter topic, i.e., the *how to* of using MATLAB features. Set B problems will center on the use of the MATLAB features of the chapter, i.e., Set B will emphasize problem solving.

This chapter though has only one problem set. In the following problem set, each problem will exercise your critical problem-solving skills. You should not take these problems to be isolated, individual problems. Rather, use the framework you have learned in this chapter and apply it to each problem.

Your challenge is to solve the problems below using the useful parts of the framework. Not each facet of the framework will be used in every problem.

Problem 2-1

For the following two situations, search for the surface area and volume of appropriate solid shapes on the Internet.

Write down your assumptions. When applicable, search the Internet and verify your results.

- The surface to volume ratio of the earth is given as $7.5753 \times 10^{-4} \, \text{mi}^{-1}$. Determine an approximate diameter for the earth. Make a sketch to illustrate the physical situation.
- The surface to volume ratio of a cylindrical silo is given as $0.2 \, \text{ft}^{-1}$. Determine the radius of the silo. List your assumptions.

Problem 2-2

In nature, in most cases more organisms are born in a population than can possibly survive. And for a very good reason: Uninhibited growth of a population is exponential. If the succeeding generations of an organism all survived, the world would be filled completely with only that organism.

For instance, if the number of organisms in a population doubles every generation, the population at the end of Nth life cycle of reproduction is 2^N. Although that does not sound like it should result in an imposing number, consider the following. After 20 generations, the number of individuals in the population would be 1,048,576. Big, but not HUGE. Now consider how many individuals there would be after 300 generations:

$$2.037035976334486 \times 10^{90}$$

The number of individuals in the population after 300 generations would be larger (substantially) than the number of elementary particles in the universe. Therefore, such unabated growth of a population will not happen in nature. Food resources or other factors in the ecosystem of the population will prevent unchecked growth.

Take a specific example. The diagram below shows a Salmonella bacteria. A Salmonella bacteria reproduces every 20 minutes. Assuming that the mass of a single Salmonella bacterium is 10^{-12} grams, how long would it take for a single Salmonella bacteria to produce a mass of 4 kilograms?

First, guess how many minutes/hours/days it would take and then find an approximate solution with a calculator using trial and error. Remember to state your assumptions.

(diagram from:
   http://www.schoolscience.co.uk/content/4/biology/sgm/sgmbugs2.html)

Problem 2-3

We want to design cylindrical bean cans to optimize transportation of the product. The cans are loaded in boxes, and the boxes are stacked on top of each other. However, there is a risk of crushing the cans at the bottom if the boxes are stacked too high.

The cans are made of aluminum plates of 2 mm thickness. Each cylindrically shaped can has a volume of 75 cm^3, weighs 130 grams, and can carry a load of 3 kgf. (Go to the Web and find out what a "kgf" is.) The truck container depth is 2 meters and we would like to utilize the entire container to optimize the transportation time and cost.

Find the optimum dimensions for the bean cans. How much aluminum is used for each can?

Problem 2-4

According to the U.S. National Highway Traffic Safety Administration, driver fatigue results in 240,000 accidents in the U.S. per year. Suggest remedies for this problem. Formulate this problem using the following:

- variables that affect the fatigue-related accidents
- potential regulations and preventive measures
- constraints on the applicability of proposed solutions
- assumptions that underpin your problem formulation

Problem 2-5

A chemical reaction takes place in a closed (and sealed) reaction chamber. The chamber is initially filled with reactants only. The amount of the reactants and the total weight of the chamber are shown in the figure below.

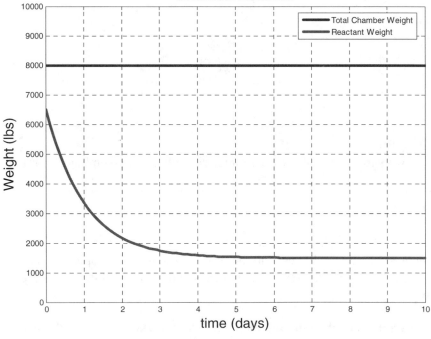

Fig. P2-5. Reaction Chamber graph.

What is the total amount of product produced in the reaction chamber? Explain how you got your answer.

Remember to note all assumptions.

Problem 2.6

A canal lock is a mechanism that lifts or lowers boats from one water pool to another. For instance, in dams, where there is a water level difference between the upstream and downstream sides, canal locks are used to transport boats from one side of the dam to the other.

You can find pictures of one set of very large canal locks that support the Panama Canal at http://www.pancanal.com/eng/photo/index.html. Click on "Canal Pictures" and click on the map location "Miraflores Locks."

More simple locks than those at the Panama Canal can be built from a series of oak and elm gates that facilitate controlled release of water to raise or lower the water level. When a boat going downstream approaches the first chamber of the lock, the water is released into the chamber to equalize the water levels. After the boat enters the chamber the water is released from the chamber to equalize it with the water level on the other side of the lock. St. Lambert Lock in Montreal is one of the largest locks of the St. Lawrence Seaway; it connects the Atlantic Ocean to the Great Lakes.

Google "St. Lambert Lock" and click on "How canal locks work" to understand the upstream and downstream operations of canal locks.

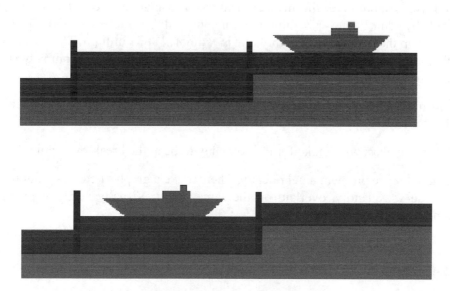

Fig. P2-6. Locks diagram.

The force applied by the water on the gates can be calculated with the following expression:

$$F = \frac{1}{2}\rho g h^2 w$$

ρ is the water density, g is the gravitational acceleration, h is the height of water, and w is the width of the gate.

For a new canal lock we are designing, the difference in upstream and downstream water levels is 4 meters. What is the water force applied at the bottom of the gate per feet of width?

Oak Thickness (in)	Strength (lbs)	Price ($)
35	155,000	2,700
42	181,000	3,116
48	212,000	4,563
61	263,000	7,811

Looking at the table above, which type of oak would you choose to use at the gate?

Problem 2-7

At Edsel Automotive, the management team is planning to expand one of its plants by adding a new assembly line for sport utility vehicles (SUVs). The cost of setting up the new SUV assembly line is estimated at $7 million. The cost of manufacturing (raw materials, labor, etc.) an SUV is $36,000 and the company is planning to sell each SUV for $38,500.

Industrial engineers at Edsel say that the new assembly line will roll off nine SUVs a day.

How many days will it take for the company to be at the break-even point?

First, solve the problem algebraically. Then, make a graph of cost vs. quantity and revenue vs. quantity, overlaid on the same graph. Visually verify your answer using your graph.

Problem 2-8

Solve Problem 2-7 assuming that the unit price for an SUV drops as the quantity manufactured increases. The fixed and manufacturing costs remain the same, but the production quantity Q and the selling price P are related by the following equation:

$$Q(P) \ = \ 7700 - 0.15P$$

Give a reasonable explanation for the variation of selling quantity depending on the price.

Make a graph of revenue vs. quantity and cost vs. quantity and visually determine the profitable range for the production.

Use a calculator and try different values for Q in this range to find the number of SUVs to produce for maximum profit.

Problem 2.9

The layout of the train tracks between Chicago and Detroit are sketched below:

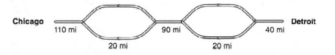

You start a stopwatch at time zero (t=0) when one train leaves Chicago going to Detroit, and at the same time (t=0) another train leaves Detroit headed for Chicago. The "pockets" on the train tracks are used for trains to pass each other.

The maximum speed for these trains is 60 m.p.h. (*Google* "TGV" for a contrast in modern train speeds.) For each path, find the optimum speed for both trains to reach their destination in the shortest time. The AMTRAK management tells you that passengers get bored and complain if a train comes to a full stop, so the goal is to keep trains moving by slowing down as necessary.

Assume that trains depart from both cities hourly. Schedule the speeds for trains from Chicago and Detroit for each path so the trains run continuously with no crashes.

Chapter 3

MATLAB Basics: Scalars

In this chapter we start using MATLAB as a tool in technical problem solving. MATLAB is one of the best-developed computer tools for supporting technical work. The basic core of MATLAB consists of a large number of built-in functions and a line-oriented Command Window in which a user executes commands that direct MATLAB to perform computational operations. On one level, MATLAB can be used as a very sophisticated calculator.

But that is just the start of what can be done with MATLAB. There is a full featured programming language that is available to the MATLAB user for writing, executing, saving, and sharing complete sequences of MATLAB commands to solve complex problems. There is in addition a very versatile plotting facility that enables simple-to-create data charts that can be indispensable for solving some types of technical problems. Further, there are tools in MATLAB that allow visual programming and simulation of dynamic systems. Finally (at least in terms of this list) there is a wealth of toolsets that are available in MATLAB ranging from a signal analysis toolset, to a toolset for sophisticated statistical operations, to a toolset for writing artificial intelligence programs, to a toolset for advanced analysis of business systems, etc. You can find a complete listing of MATLAB components and toolboxes at http://www.mathworks.com.

With so much to explore, not all of the features and capabilities of MATLAB will be covered in this textbook. Rather, the purpose here is to introduce you to MATLAB, enable you to learn a beginning set of capabilities, and empower you to learn more of MATLAB's capabilities when you need them. This journey starts with learning to use MATLAB as a sophisticated scientific calculator for simple numerical computations (scalars), and that is the topic for this chapter.

3-1   The First Time You Open MATLAB

Figure 3-1 shows a view of MATLAB as it appears when you first launch it.[1] You might notice first that this view shows MATLAB greeting me: "Welcome Jon." MAT-LAB is extensible so it is easy to change as you see fit. Your version will start with a general greeting. In a later section you will learn how you can customize the MAT-LAB environment to your own individual tastes.

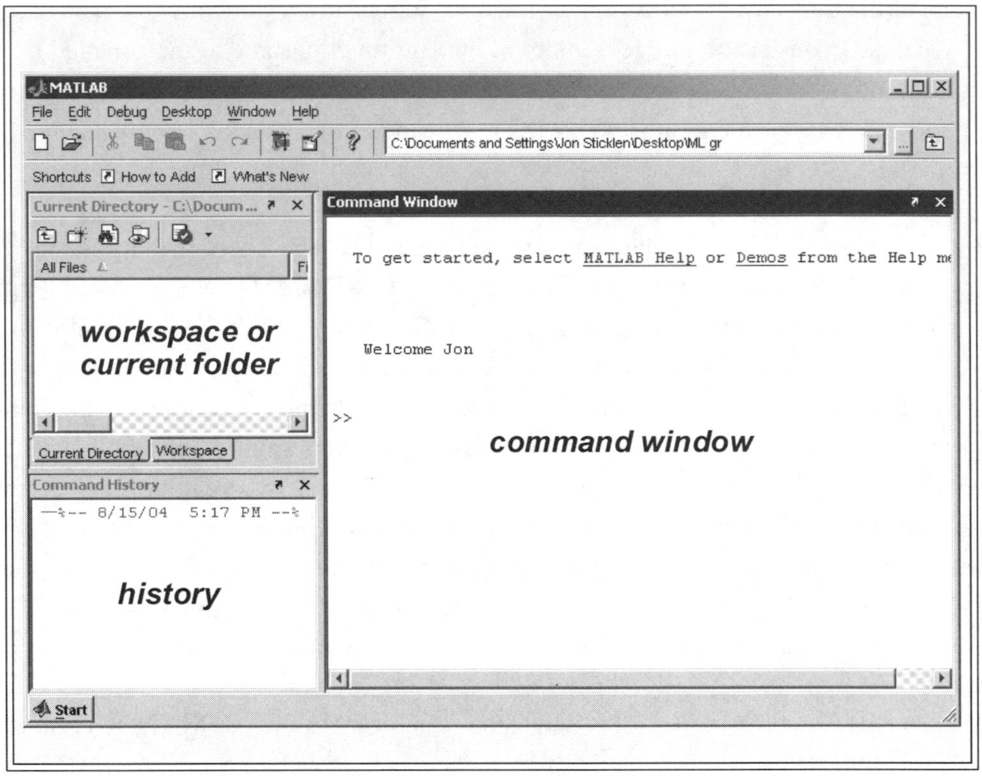

Figure 3-1: MATLAB as it Appears Immediately After Launch

Using MATLAB has a steeper learning curve than you may have experienced before. One of the keys to mastering MATLAB is to learn, and learn well, how to use all of the major components in the MATLAB environment. These components include the following:

---

1. The description in this section, and in this book generally is in detail for MATLAB release 7.0. Earlier versions of MATLAB will come up slightly differently, that is with different windows showing and different organization of the windows.

1. A Command window in which you enter commands to MAT-LAB for execution
2. A Workspace window that lists all the variables that you have set by Command Window execution
3. A Command History window that records each command you execute in the Command Window
4. A Current Folder window that lists all the files (data files and program files) that are directly available for MATLAB use
5. An Editor window that is used for implementing MATLAB programs
6. An Array Editor window that allows direct editing of MAT-LAB arrays
7. A Help window from which you can receive help about any aspect of MATLAB
8. A window from which you can set preferences for how MATLAB displays results, what font is used for display, and many other aspects of how MATLAB looks to you, and how it operates for you

The first four of these windows are available when you launch MATLAB from your PC either by double-clicking the MATLAB icon (if you have one on your desktop), or by going to your START button and selecting Programs, then finding MATLAB and selecting it. We are going to discuss windows 1, 2, 3, 4, 6, and 8 in this chapter.

First, let us set the way we want MATLAB to display numbers using the eighth window in the list above, then one used for setting MATLAB Preferences. Pull down the **File** menu, and select **Preferences**. Many preferences can be set to allow you to customize MATLAB. In the left window pane, click on **Command Window** as shown in Figure 3-2. Then, also as shown in the figure, click on the pull-down for Numeric format. The format **short** is the format that MATLAB will show selected. For now, **short** will be the selected value. The **short** format for numbers shows numbers to only four decimal-point accuracy, but for most of our initial purposes that is all we need, and it makes MATLAB output easier to read. As we progress through this book, you will learn how to use other MATLAB preference settings. Better yet, explore the Preferences possibilities yourself.

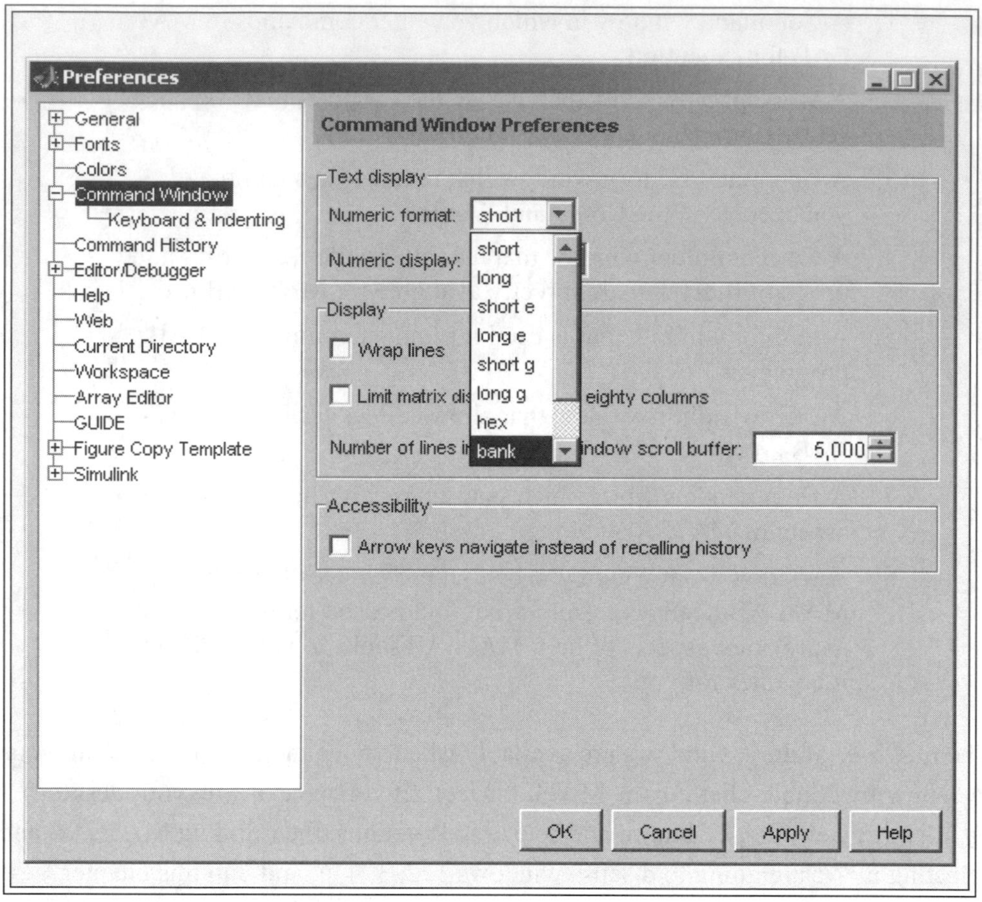

Figure 3-2: Setting MATLAB Preferences to Display an Easy-to-Read Number Style

Now, type the following line into the MATLAB Command Window:

        >> matpie = 3.1416

Then press **ENTER**.

MATLAB answers you by letting you know that the variable **matpie** is now set to 3.1416. Click the *Workspace* button below the window marked **workspace or current folder** as shown in Figure 3-1. Notice that an entry is in the Workspace window of variables, the variable **matpie** that you created. You can see the value of **matpie** you just entered.

In the Command Window, type the following lines, pressing **ENTER** after each.

        >> diameter = 2
        >> area = matpie * (diameter / 2)^2

You have created two more variables, **diameter** and **area**, and you have used MATLAB to calculate the area of a circle with diameter of 2, for which you find the area to be 3.1416 (units are not shown). Again, notice that the new variables are also listed in the Workspace window. Look back at the lines of code entered in the MATLAB Command Window. Notice the MATLAB *operator symbols* for multiply, divide, and to-the-exponent. We will discuss MATLAB operators shortly.

You may wonder at the spelling of the variable we named **matpie**. As you probably guessed, this variable was to hold the value of π. So why not name the variable "pi?" It's because MATLAB already has a *predefined* variable named **pi**. (We could have used the predefined **pi**.)

Enter the following line to the Command Window and press **ENTER**.

>> **pi**

The variable **pi** is an example of a predefined variable setting in MATLAB.

Your MATLAB windows on your computer monitor should be similar to those shown in Figure 3-3.

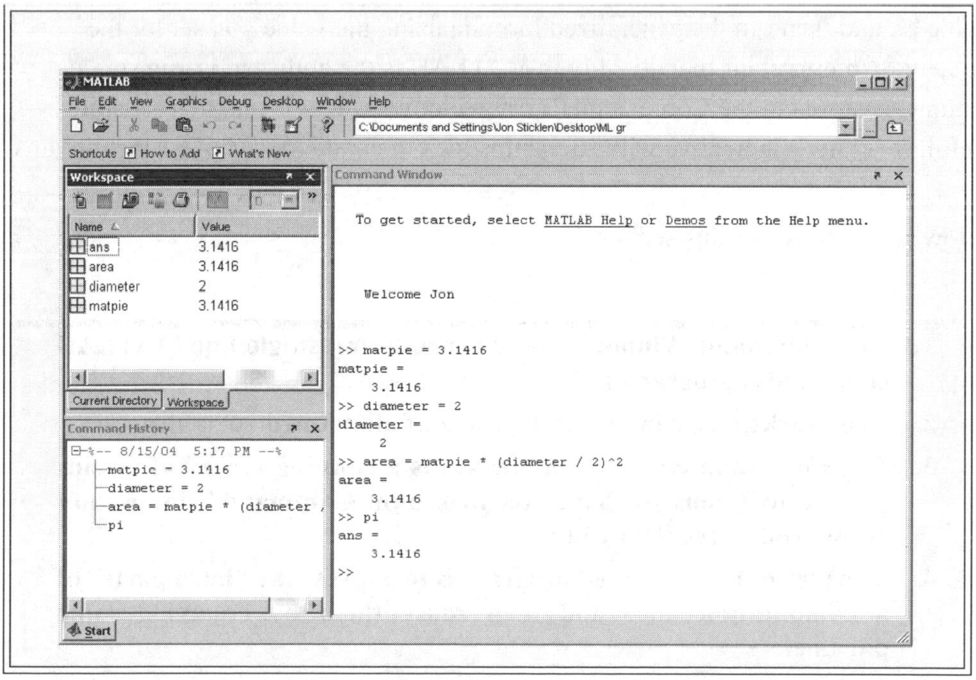

Figure 3-3: MATLAB Windows After Doing Calculations for Circle Area

What was the purpose of typing the value of **matpie** and **diameter** to be executed in the Command Window prior to typing in the formula for a circle? You could have just as easily typed in the following line:

```
>> area = 3.1416 * (2 / 2)^2
```

The answer to that question is easy and yet profound. To understand, type the following line to the Command Window:

```
>> diameter = 3
```

Note the new item that appears at the bottom of the Command History window. In the Command History window, double left-click the line that you typed in previously to calculate the area of the circle. The line is entered into the Command Window and executed as though you typed it in and pressed ENTER. You just calculated the area of a circle with a diameter of three. The answer displayed in the Command Window is 7.0686.

You have just created your first MATLAB program! It is an admittedly simple program, one you could have solved with a calculator or with a pencil and paper. Nonetheless, it qualifies as a computer program. You have used the facilities of MATLAB to save a generalized formula for the area of circle in terms of its diameter (saved automatically for you in the Command History window), set the value of the diameter, and then run the generalized formula using the value you set for the diameter. An important functionality in MATLAB is the ability to re-run single commands stored in the Command History window. You will find this capability useful. In a later chapter, we will extend this idea to create true MATLAB programs.

Below is a synopsis of this section.

---

1. **The Command Window is used for entering single line MATLAB commands (programs).**

2. **The Workspace window keeps track of all defined variables.**

3. **The Command History window keeps a running record of all single line programs you have executed in the Command Window and allows you to recall any of them.**

4. **The idea of a generalized program is to express the "hard parts" of a computation once and then reuse them by changing some parameters.**

Synopsis for Section 3-1

3-2   MATLAB as a Calculator for Scalars

In the section above, you learned how to bring up the MATLAB computational environment on your computer, the uses for some of the primary windows in the MATLAB environment, how to perform simple computations with MATLAB, and how to write a simple MATLAB program by making use of the Command History window in MATLAB. In the sections below, you will systematically expand your capabilities. Keep in mind the purpose of this section—we have a LONG way to go before you will be comfortable with MATLAB. The first goal on the path toward proficient use of MATLAB is to build up a basic MATLAB vocabulary and a working knowledge of basic MATLAB operations. In the current chapter, we deal only with MATLAB operations on scalars.

A **scalar** is what you have simply called a "number" through most of your training in mathematics. It is a single entity that cannot be broken down further, and is a magnitude on a standard scale. That is, the scalar 3.5 is a number on a standard scale of other numbers. 3.5 is bigger than 3.4 and smaller than 3.6 on the real number scale. In a mathematical sense, scalars can be integers, rational numbers, or irrational numbers. MATLAB handles all standard numbers from mathematics and goes further with scalars to handle infinitely small or infinitely large numbers. In this text, we are not going to deal with complex numbers, but you should remember that MATLAB will handle complex numbers in case you need that capability for purposes after you complete this book.

Why all the fuss about defining something (scalars) that you already knew about, though simply called "numbers"? The reason is that MATLAB goes beyond simple numbers to handle groups of simple numbers all in a line (called **vectors** in MATLAB), table-like groupings of numbers (called **arrays** in MATLAB), multidimensional arrays of numbers, and other more specialized representations of numbers and characters. In due course we will get to many of these other forms beyond simple scalars, and the operations that go with each. But we start with the simplest form—scalars. We give this form the name **scalar** to be sure we do not cause confusion with other forms we will be dealing with later.

There is one cautionary note to add about the term **scalar**. In science, mathematics, and engineering, the term **scalar** means *simple number* in the sense we have used above. But it is usually used to specify some quantity, in contrast to being a vector (or tensor). In the MATLAB environment, **scalar** is used to mean a simple number, too, but it is usually in contrast to being an **array**. Keep this in mind. When you hear in your physics class, for example, that velocity is not a **scalar**, it is because velocity in physics is a physics **vector**. When we say here in the context of MATLAB that x is not a **scalar**, it is because the variable x is an **array** (or some other non-scalar).

3-3   Fetching and Setting Scalar Variables

In this and succeeding sections you will develop basic skills for using MATLAB as a sophisticated calculator for technical problem solving. The topic for this section is MATLAB variables. Before going on to the specifics for MATLAB variables though, you need a good mental model of what a variable is in *any* computer environment.

3-3.1   A Mental Model for Computer Variables

A useful way to think about computer variables is that they are **named containers**. Once one of these containers (variables) exists, we can perform two types of operations:

1.   we can look at the value held in the container, and
2.   we can set the value held in the container.

Looking at the value in the container is an operation called *fetching*. Once we know a container's (variable's) name, fetching it in MATLAB means just writing that name. If a container (variable) is named **x**, and the container currently holds the value 22, then we can fetch its value in MATLAB just by entering in the Command Window:

        >> x

MATLAB dutifully answers the following:

        x =
            22

We can perform whatever operations we want with the value of **x** (the contents of the container named x), like adding 45 to it:

        >> x + 45

MATLAB performs the indicated operation, and gives us the answer:

```
ans =

   67
```

Note that when we enter the name of a variable in the Command Window, MATLAB answers with a line having the name of the variable, the equal sign, and on a second line the value contained in the variable. But, when we enter an *expression* (e.g., **x +** **45**), MATLAB answers with **ans =** on one line, then the value of the indicated computation on a second line. **ans** is another predefined variable in MATLAB. It is reserved for the result (the answer) of a computation when no variable is stated to hold the result.

To set the current value of a variable (contents of a container), you use the MATLAB *assignment operator.* You already saw examples of variable assignments in Section 3-1 above, but our purpose here is to talk about assignments more formally, and in terms of the container model of variables that we are developing.

The equal sign is the *assignment operator* in MATLAB: to the left of the equal sign is the name of the variable; to the right of the equal sign is the value to which you are setting this variable. This is where the *container model* of computer variables comes in handy. In terms of the container model, assigning the variable **x** the value of 22 is shown in the sketch in Figure 3-4.

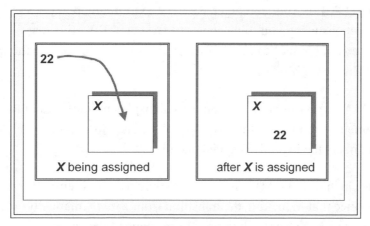

Figure 3-4: Container Model View for Variable *x*
Being Set to 22

Why all the seemingly endless fuss about a simple operation? It's because through all the years of math classes, particularly when you studied algebra, you have come to understand a mathematical relation with an equal sign differently than the meaning of the assignment operator in MATLAB. To understand this distinction, consider the following operation in MATLAB:

```
>> x = x + 1
```

In algebra, the statement above would be invalid because you could subtract *x* from both sides of the "equation" and end up with **0 = 1**, which is obviously false. An algebraic equation asserts that whatever is on the left side of the equal sign is *equal to* whatever is on the right of the equal sign. Many conceptual operations in algebra then amount to "doing the same thing to both sides of the equation." In contrast, in MATLAB, the equal sign is the symbol for the *assignment operator*. The perfectly valid MATLAB operation **x = x + 1** conceptually means to do the following steps:

1. Look in the container named **x**.
2. Get a copy of the value stored in that container.
3. Add 1 to that value.
4. Place the sum back into the container named **x**.

NOTE: Be certain you understand and remember the distinction between "equal" in algebra versus "assign" in MATLAB. Many beginning students confuse the difference between algebra *equals* and MATLAB *assignment* at the beginning and have a difficult time straightening out the confusion later.

3-3.2   Setting Variables for the Hay Bale Problem (from Section 2-2)

Bring up MATLAB on your computer. As we step through operations in MATLAB, do those operations using your own copy of MATLAB. Effective learning of MATLAB, or any computer application, must include your active participation; you must *use* MATLAB to learn MATLAB.

Change the MATLAB Preference for numerical display to **bank**. Refer to Figure 3-2 if you don't remember how to do this. The bank numerical format in MATLAB results in displayed numbers having only two decimal places.

The variables in the MATLAB solution for the hay bale problem (HBP) are listed in Table 3-1. Notice that in making the transition from a mathematically-oriented sketch of the solution to a MATLAB solution we have changed the variable names to be more easily understood. Good computer variable name selection is usually a trade-off

**Table 3-1: MATLAB Variables Used for Solution of
The Hay Bale Problem**

MATLAB variable name	meaning of variable in context of the Hay Bale Problem
accelGravity	The acceleration due to gravity. A constant for the Hay Bale Problem. Equal to 32 feet per second per second.
speedInitial	The initial speed given to the hay bale by the catapult. Units are feet per second.
speedInitial_Vert	The component of the initial speed in the vertical direction.
speedInitial_Horz	The component of the initial speed in the horizontal direction.
phi	The initial firing angle - the angle at which the catapult launches the hay bale, in degrees, from the vertical.
horzDist	The horizontal distance the hay bale travels given phi and speedInitial.
heightMax	The maximum height the hale bale reaches.
phiMax	The value of phi that will maximizes horzDist.
horzDistMax	The value of horzDist that is maximal.

between having names that are short (and hence easy to type) and variables that are easily understood. Using single character variable names is usually a bad idea in computer solutions because they convey almost no meaning as to the values these variables hold. Unless you are told which variable names to use, it is up to you to select reasonable names, and that means names that help you or anyone else understand your computer solutions.

Legal variable names in MATLAB do not start with a number, and they may not include the symbol for any MATLAB operator (see the following information on MATLAB operators). Although **pi** is a legal variable name in MATLAB, if you set **pi** to some value, then you overlay the built-in constant value of **pi**. The same is true for any of the MATLAB built-in constants. Those are the only rules about naming variables you need to worry about. Naming variables is more a matter of style than a matter of absolute rules. Over time, you will learn by experience how to name variables well and to meet that trade-off between shortness of names and understandability.

Open MATLAB if you do not have it running. If you are opening MATLAB fresh, then execute the following:

```
>> x = 22
```

Be sure you have the Command History window and the Workspace window displayed. Observe the Workspace window, then execute the following line.

```
>> clear
```

Observe the Workspace window again. The MATLAB command **clear** deletes all current variables from the Workspace. Before starting to work on a new problem, you should always clear away any variables that you have been using so that you are starting anew. In terms of the container model for variables, **clear** means to throw away the container. Doing this will prevent you from being confused should a variable from an earlier problem have the same name as a variable you want to use in your current problem.

Now we are ready to set up the variables in Table 3-1. Execute the following lines in the Command Window:

```
>> accelGravity = 32;    %    units: ft/sec/sec
>> speedInitial = 50;    %    units: ft/sec
>> phi = 10;             %    units: degrees
```

Each one of these lines is an application of the assignment operator. Observe again the Workspace window. Three variables are listed, which are precisely the variables you just created.

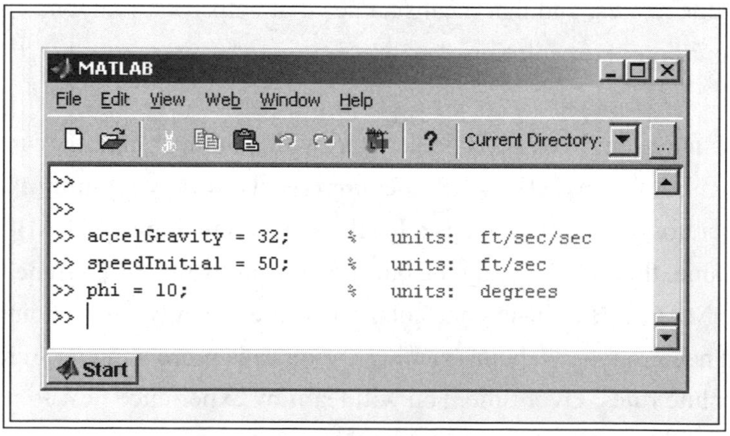

Figure 3-5: Snapshot of Command Window:
Semicolon Prevents Display of Results

We have introduced in these three assignment operations two new MATLAB facets. The first is the MATLAB symbol used to indicate a comment. We often want to write notes into lines of code to remind us of something. The MATLAB comment serves this purpose. In a line of MATLAB code, if you type a percent sign, anything after the percent sign is ignored by MATLAB as it runs your program. Whatever comes after the percent sign is for your reference only. We used comments above to indicate the units of the three variables we created.

The second new MATLAB facet we introduced is the MATLAB symbol used to execute a code line in the Command Window with no resultant answer being displayed. If you end a command with a semicolon, there will be no answer back from MATLAB indicating the result of executing the command. To make this difference explicit, first examine your MATLAB Command Window. It should look like that shown in Figure 3-5. If the commands had not included the semicolon ending, then the results in your Command Window would have looked like that shown in Figure 3-6. Be sure you understand the difference between using and not using the semicolon at the end of a code line. When would you use a semicolon at the end of a MATLAB command and when would you not? (Do not answer *when you want the answer shown.* That is not enough. Under what circumstances would you like to have the result of a MATLAB command shown to you?)

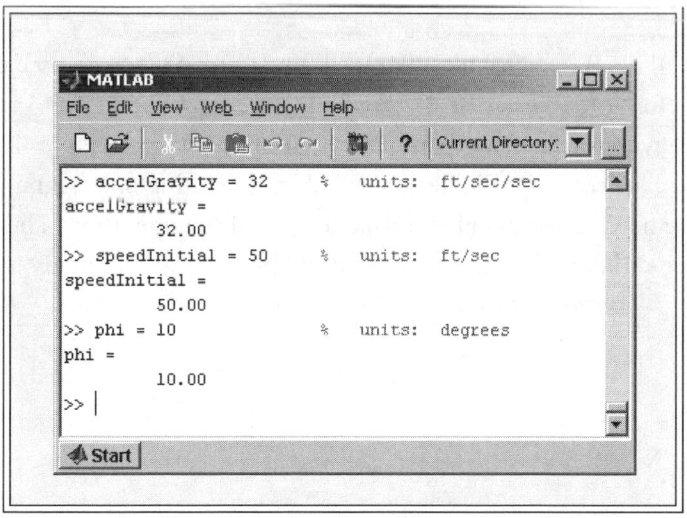

Figure 3-6: Snapshot of Command Window:
Lack of Semicolon Enables Display of Results

Your MATLAB Command History window should now look similar that shown in Figure 3-7. If you want to change the value of **phi** but have forgotten the units (degrees or radians), then a look back at your initial set-up for the variables in the Command History window will refresh your memory.[2] If you want to change the initial value of **phi**, then just copy/paste from the Command History window to the Command Window, change the value that **phi** is set to, and execute the line. Again, the Command History window is a tremendous aid in using MATLAB if you make the effort to learn its uses.

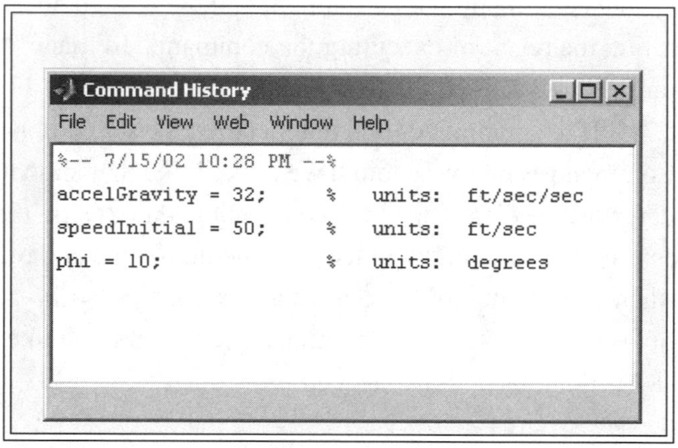

Figure 3-7: Snapshot of Command History Window

We have created the three variables that we know (**accelGravity**), or that we can assume values for, initially (**speedInitial**, **phi**). In addition to creating these variables, we have set values for them. But what of the other variables in Table 3-1? In MATLAB, we can create a variable as needed by executing an assignment operation and naming the new variable on the left-hand side of the operation. Thus, we can hold off on the other variables for the Hay Bale Problem until we are ready to assign values to them.[3]

---

2. This idea of writing notes for personal use into code becomes more significant when we get to storing programs in files for later use as opposed to our current use of the Command History window.

3. Many computer languages demand that you *declare* a computer variable before you use it. MATLAB has no such restriction.

3-4   MATLAB Built-in Functions, Operators, and Expressions

In the last section, you learned about MATLAB scalar variables. Your next step is to learn how to build commands to set scalar variables. A good deal of what you learn about operations to build computations in this section on scalar operations can be generalized to operations on vectors and matrices, so this material is important. As you work through this section, be sure to open MATLAB and perform the operations as you are reading about them.

The specific goal for this section is for you to understand the ideas of MATLAB built-in functions, operators, and expressions. We will defer returning to the Hay Bale Problem until we have covered all the material needed to completely solve the problem, and that means we need to get through the material on arrays that come in a later chapter.

3-4.1   Built-in Functions

A *built-in function* is a function that MATLAB comes with when you launch it. There are hundreds of built-in functions in the standard student edition of MATLAB, and a very large number of additional built-in functions in the many available toolboxes. All of these built-in functions are available from the MATLAB Command Window. Most MATLAB built-in functions are named to be closely coincident with names you already know for these functions. When we set up the Hay Bale Problem in the last chapter, we used liberal doses of trigonometry. One of the trig functions used was *sine*. In MATLAB, the *sine* built-in function is written just as we wrote it in Equation 2-5: `sin`.

The concept of built-in MATLAB functions is simple as long as you keep in mind the general sense of what a computational function is. More tedious is remembering the names of MATLAB functions you will be using. Later we will discuss using the extensive MATLAB Help facilities to find the names of MATLAB functions. Until the Help facility is described in Chapter 6, we will introduce each new built-in MATLAB function as needed.

A few often-used MATLAB built-in functions are shown in Table 3-2. These built-in functions represent the tip of the iceberg in relation to all the built-in functions that MATLAB offers.

Table 3-2: Sample of Often-Used Built-in Functions

description	MATLAB syntax
sine of x	sin(x)
tangent of x	tan(x)
log base 10 of x	log10(x)
natural log of x	log(x)
e^x	exp(x)

3-4.1.1 MATLAB Operators

You already know one MATLAB operator: the assignment operator that was introduced in Section 3-3.1. In general, MATLAB operators can be viewed as a special instance of built-in functions. The difference is in the way you write the line of MATLAB code that invokes the operator. For a function named f that takes two arguments, arg_1 and arg_2, you would write the following:

```
>> f(arg1,arg2)
```

If f were an operator, then there would be a more direct way to write the invocation. If, for example, f were an operator that added two numbers, then this is the MATLAB manner of invoking the **plus** operator:

```
>> arg1 + arg2
```

A MATLAB operator is an often-used part of MATLAB, so often that it is elevated into a subclass of all built-in functions and is given its own symbols for invocation. The bottom line is that the special subclass of often-used MATLAB function, the operators, is defined to make your job of writing MATLAB code easier. Instead of being forced to write

```
>> plus(3,5)
```

it is much more convenient to write[4]

```
>> 3 + 5
```

Both of the two options above — the built-in function form and the operator form of the plus operation — work in MATLAB.

---

4. If you are interested in issues of computer science, then ask, "What is it that makes the second form more convenient?" User convenience, called user friendliness, is a driving force in computer science, not only for the design of programming environments like MATLAB, but more generally in all computer applications.

We are going to use four classes of operators:

1. **The assignment operator**: The assignment operator is in a class by itself.

2. **Arithmetic operators**: This class of operators enables normal operations of arithmetic on two numbers.

3. **Relational operators:** This class of operators enables comparing two numbers.

4. **Logic operators:** This class of operators enables combining the results of one or more relational tests.

3-4.1.1.1   Arithmetic Operators

Table 3-3 summarizes the arithmetic operators in MATLAB. These operators have meanings with which you are familiar.[5]

Table 3-3: Arithmetic Operator Class

operator name	operator symbol	example
unary plus	+	+ x
unary minus	−	- x
addition	+	x + y
subtraction	−	a - 22
multiplication	*	pi * r
division	/	circumference / diameter
exponentiation	^	radius ^ 2

For example, suppose that you want to find the fourth root of 256. One way to perform that operation in MATLAB, using the exponentiation operator, is as follows:

```
>> 256 ^ 0.25
ans =
        4.00
```

---

5. One exception may be the unary minus operator. The effect of unary minus on a number or numerical variable is to flip the sign of its argument.

Arithmetic operators take numbers or numerical variables as arguments and produce numbers.

3-4.1.1.2   Relational Operators

Table 3-4 summarizes the relational operator class in MATLAB. The ideas backing these operators are most likely not new to you. For example, suppose that we compute the average of a sequence of two numbers, {75, 92}, and then ask the question if the average is less than 80. Copy the code just below into your MATLAB Command Window one line at a time and execute each. Notice that all these commands end with a semicolon, except the last. The reason? Because the first four lines of code should have a clear meaning to you now, though the last line may not, we display the result of the last code line:

```
>> x1 = 75;
>> x2 = 92;
>> sum = x1 + x2;
>> average = sum / 2;
>> classAverageTest = (average > 75)
```

Your Command Window and Command History window should appear similar to that shown in Figure 3-8 after execution of the five lines of code.[6] Double left-click on the entry for the variable named **average** in the Workspace window to see what its numerical value is set to. Noting that **average** now contains the value 83.5, the test

Table 3-4: Relational Operator Class

operator name	operator symbol	example
LESS THAN	<	x < y
LESS THAN, EQUAL TO	<=	a <= 22
EQUAL TO	==	x == 100
NOT EQUAL TO	~=	x ~= 32
GREATER THAN, EQUAL TO	>=	pi >= 3
GREATER THAN	>	classAverage > 75

to determine if the value of **average** is greater than 75 has a clear result. We expect the answer to be Yes or No, and that leads us to discussing how MATLAB deals with *logical variables* and *logic values*.

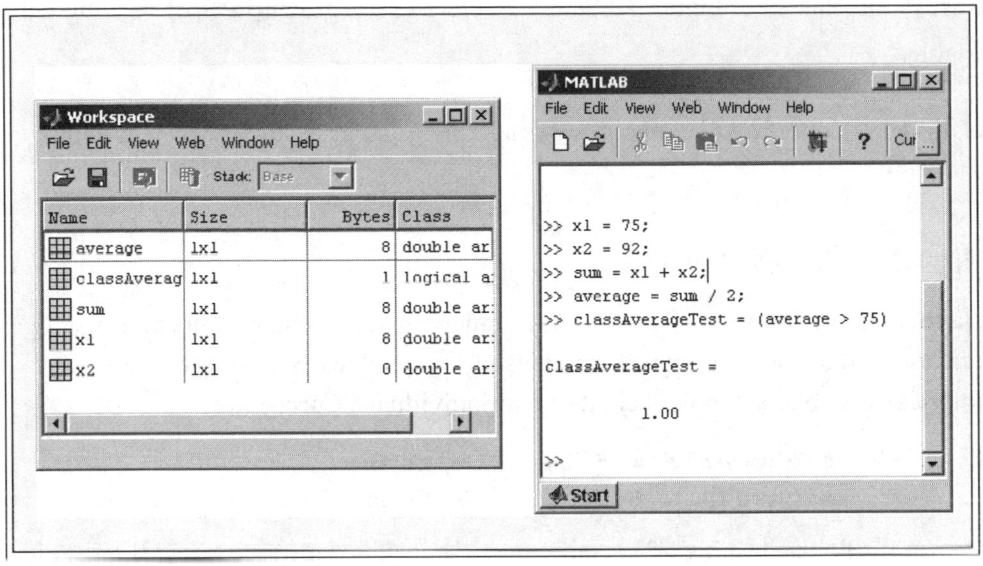

**Figure 3-8: Simple Sequence of MATLAB Commands
to Average Two Numbers
and Test Whether the Result is Greater than 75**

There are an infinite number of numerical values,[7] but only two possible logic values exist: **TRUE** and **FALSE**.[8] In MATLAB, logic values are assigned to numbers. The number zero viewed as a logic value is **FALSE**. *Any other number viewed as a logic value is* **TRUE**. This manner in which MATLAB defines logic values is unique among computer languages and is a point that beginning MATLAB users often miss. You can understand a logic variable in MATLAB by thinking of it as any variable that contains a logic value.

---

6. If you are watching carefully, you might notice that in Figure 3-8 the Command Window is separated from the Command History window though in your version of MATLAB they are connected. This separation is one of the customizing features of MATLAB. We will discuss this later in the text, but if you want to experiment now, pull down the View menu in your Command Window and experiment with *desktop layout*.

7. For those with a strong mathematical bent, there is an interesting story here. The number of all integers is infinite. Yet between any two integers there are an infinite number of rational numbers. Thus, intuitively, the number of all rational numbers is a larger infinity than the number of all integers. Thus, there are different levels and different sizes of infinity! Results from mathematics support this intuition.

8. Strictly speaking, only two logic values exist in *Boolean logic*. Other types of logic have more than two values.

But if a variable **x** contains the number 1, then is that variable a numerical variable or a logical variable? Read the previous paragraph carefully. Note the twice-used phrase *viewed as a logic value*. The answer to the question is: it depends. If **x** is used as a logical variable, then its value is either a logic value, or its value is converted to a logic value. In the context of Figure 3-8, the variable **classAverageTest** is a logical variable.[9]

Relational operators take numbers or numerical variables as arguments and output logic values.

3-4.1.1.3   Logic Operators

Relational operators are used to compare numerical variables or numbers. Logic operators are used to combine logical variables or logic values. Suppose we have two numerical variables set to the height of two individuals, Carla and Jim:

```
>> heightCarla = 5.5;      % feet
>> heightJim = 6.5;        % feet
```

Suppose further we test if Carla is taller than six feet, and also we test if Jim is taller than six feet:

```
>> carlaTallerSixFeet = (heightCarla > 6);
>> jimTallerSixFeet = (heightJim > 6);
```

Finally, suppose we want to ask if Carla and Jim are taller than six feet. We know if Carla is shorter than six feet (the value FALSE is in **carlaTallerSixFeet**), and we know if Jim is taller than six feet (the value **TRUE** is in **jimTallerSixFeet**). We can combine the two facts to answer the question: Are both Carla and Jim taller than six feet? We use the logic function **AND**. The operator symbol for **AND** is **&**.

```
>> bothTaller = carlaTallerSixFeet & jimTallerSixFeet
bothTaller =
             0
```

bothTaller will be **TRUE** only if Carla is taller than six feet AND Jim is taller than six feet. Suppose we wanted to ask if Carla OR Jim is taller than six feet. We can use the MATLAB function **OR**. The operator version of **OR** uses the symbol |. This symbol is found on most keyboards on the same key as the backslash symbol.

---

9. A full discussion of this paragraph is a topic for a more advanced class in computer science. The key in the paragraph is the phrase "viewed as." The ability of a function to work on multiple types of data values is called "operator overloading." To talk about operator overloading in a modern context requires understanding object oriented programming and is beyond the scope of this textbook.

```
>> oneTaller = carlaTallerSixFeet | jimTallerSixFeet
oneTaller =
          1.00
```

oneTaller will be **TRUE** if EITHER Carla is taller than six feet OR Jim is taller than six feet, or if both are taller.

There is one more logic operator that MATLAB offers, one that flips a logic value: If the argument is **TRUE**, the result is **FALSE**, and if the argument is **FALSE**, the result is **TRUE**. The MATLAB logic function that performs this flip is **NOT**, and the operator symbol for it is ~:

```
>> flipCarlaTallerSixFeet = ~ carlaTallerSixFeet
flipCarlaTallerSixFeet =
          1.00
```

The operators for **AND** and **OR** combine two logical arguments, but the operator for **NOT** takes just one argument. The arguments for any of the logic operators are logic values or variables, and the result is a logic value. The MATLAB logic operators are summarized in Table 3-5.

Table 3 5: Logic Operator Class

operator name	operator symbol	example
combine AND	&	x & y
combine OR	|	a | b
flip NOT	~	~ z

3-4.2   Expressions and Rules for Forming Expressions

A MATLAB *expression* is the construct formed when we bring together a number of variables, constants, or expressions for a purpose. That purpose is to perform some computation. Expressions can be thought of as a blueprint for computation.

For example, a numerical expression for computing the sum of two numerical variables **x** and **y** in MATLAB is the following:

```
>> x + y;
```

An expression in MATLAB, or any programming language, is like a paragraph in English. When we want to say or write a thought, we bring together words into sentences. Sentences, in turn, can be strung together to form paragraphs. The analogy goes further; just as there are rules of grammar in English that limit how we can form sentences and paragraphs, there are rules in MATLAB for how to form expressions. Compared to the set of grammar rules in English, the number of rules for forming MATLAB expressions is much smaller. One reason why fewer rules are needed for forming MATLAB expressions is that the grammar rules are rigid in MATLAB, while the grammar rules for English are more guidelines than rigid rules.

An expression in MATLAB has a computational purpose, e.g., "to compute the sum." Sentences or paragraphs on a paper page have a purpose to communicate an idea. But just as the purpose of sentences or paragraphs is not realized until a person reads and understands the idea, the computational purpose of an expression is not realized until MATLAB evaluates the expression. The result of evaluating a MATLAB expression is a numerical or logic value.[10]

In the descriptions above of the MATLAB scalar operator classes, we have been using expressions and implicitly using rules for forming and evaluating expressions. The purpose of this section is to make the rules explicit. There are only two rules for MATLAB expressions. The first is a rule about what constitutes a legal expression. MATLAB expressions are one of the following:[11]

1. A numerical value or variable

2. A logical value or variable

3. A legal application of a MATLAB function (a special case would be the legal application of a MATLAB operator)

4. A combination of MATLAB expressions

The third possibility uses the word *legal*. A legal application of a MATLAB function means an application of the function in which the arguments types required by the function are correct. For example, the sine function requires one argument. To obtain the results expected (the value of the mathematical sine function that applies to some angle), the single argument must be in units of radians. If we evaluate the MATLAB sine function with two variables, MATLAB does not understand what we are asking,

---

10. Remember that in this section, we are only talking about scalars. More generally, the result of evaluating a MATLAB expression is any valid MATLAB data structure. It depends on the MATLAB function we are using and on the input arguments it is given.

11. In this section we are limiting ourselves to scalar data types.

and an error is produced because the expression we gave to MATLAB for evaluation was illegal. For example,

```
>> sin(pi/2, pi/8)
??? Error using ==> sin
Incorrect number of inputs.
```

Evaluating a numerical or logic value results in the numerical or logic value:

```
>> 22
ans =
          22.00
```

Evaluating a variable results in the value of the variable. In terms of our container model for variables, evaluation of a variable results in the contents of the variable:

```
>> x = 22;
>> x
x =
          22.00
```

Evaluating a MATLAB function or operator means to apply the function or operator and return the result:

```
>> sin(pi/4)
ans =
          0.71
```

Although the first three types of MATLAB expressions and the evaluation of them are straightforward, the fourth type of MATLAB expression requires more explanation.

Saying that a MATLAB expression may be a combination of other MATLAB expressions means we can form an expression out of other expressions. For example, suppose we want to write a MATLAB expression that would give us one of the roots of the quadratic equation. Mathematically we are solving the following:

$$ax^2 + bx + c = 0 \qquad \text{(EQ 3-1)}$$

This allows us to find:

$$x = \frac{-b \pm \sqrt{b^2 - 4ac}}{2a} \qquad \text{(EQ 3-2)}$$

Suppose we want the root that takes the positive sign of the plus/minus option. As a first step, we need to set values for the constants of a specific example of the quadratic equation. Suppose we are solving for $a = 1$, $b = 3$, and $c = 2$:

```
>> a = 1;
>> b = 3;
>> c = 2;
```

Suppose we concentrate first on the numerator of the right side term in Equation 3-2. Remember, we are taking the positive sign for the plus/minus option. We can form a complex expression for the numerator, an expression made of other expressions:

```
>> numerator = (-b) + ( (b^2) - ( (4*a) * c ) )^0.5
numerator =
           -2.00
```

Follow the way the parentheses in this complex expression are arranged. In each case, the most nested pairs of parentheses set off a simple expression, an expression that can be directly evaluated. At the other extreme, the expression above is an assignment operator expression. MATLAB understands then that we want to assign the variable called **numerator** some value. The value to be assigned to **numerator** is on the right-hand side of the expression above:

```
(-b) + (   (b^2)   -   ( (4*a) * c ) )^0.5
```

Evaluation in MATLAB starts at the left of this expression. Hence, the **(-b)** term is evaluated first. This term is a parenthesis pair with a simple expression inside. The result is added to the result of evaluating the second term. When MATLAB interprets the second term, it dives down into the parenthesis pairs until it reaches bottom and evaluates the simple expressions at that lowest level: **-b**, **b^2**, and **4*a**. Then MATLAB works it way back up through the parenthesis pairs evaluating each complex expression in turn (*in turn* means as indicated by the parenthesis pairs). In general, MATLAB is working by two rules in the evaluation of complex expressions:

1. Evaluate from left to right
2. Combine simple expressions in the order indicated by the placement of parenthesis pairs

One exception to the use of parenthesis pairs is that any expression in MATLAB can be enclosed by parentheses *except* an expression that is the assignment operator.

Writing all the parenthesis pairs is tedious and prone to error. A much more natural way of writing an expression for the numerator term above would be as follows:

```
>> numerator = -b + ( b^2 - 4*a*c )^0.5
numerator =
           -2.00
```

The complex expression on the right of the assignment statement is our focus: **-b+(b^2-4*a*c)^0.5**. MATLAB starts on the left side of this expression for evaluation. The **-b** term is a simple expression, i.e., a use of the numerical unary minus operator. MATLAB evaluates it to be −3. That value will be added to the next expression, but it is a complex expression: **(b^2-4*a*c)^0.5**. The parenthesis pair lets MATLAB know we mean to perform all the computations inside the parentheses before the exponentiation is done. So, we can concentrate on the complex expression inside the parentheses: **b^2-4*a*c**. If we read this complex expression left to right, we would do the following steps:

1. Square the value of **b**

2. Subtract **4** from the result of Step 1

3. Multiple the result of Step 2 by the value of **a**

4. Multiple the result of Step 3 by the value of **c**

But this would be wrong. The result produced would not be the value intended. Verify this for yourself. What we intend mathematically is the following steps:

1. Square the value of **b**

2. Multiply **4** times the value of **a**

3. Multiply the result in Step 2 times the value of **c**

4. Subtract the result of Step 3 from the result in Step 1

This last set of steps produces the result intended. Verify this for yourself.

MATLAB enables us to write simpler code, like the complex expression **b^2-4*a*c**, and obtain expected results that coincide with standard mathematics by using Rules of Precedence for the evaluation of complex expressions. The Rules of Precedence tell MATLAB the order in which to evaluate the parts of a complex expression. These are the two parts of the Rules of Precedence for MATLAB:

- Multiplication and division must be performed before addition and subtraction

- Exponentiation must be performed before multiplication and division

With these two parts of the Rules of Precedence, the second set of steps are followed by MATLAB in evaluating **b^2-4*a*c**.

The Rules of Precedence for MATLAB operators are shown in Table 3-6. This statement of the Rules of Precedence is valid only for scalars. We will need to update it when we discuss vectors and arrays. The way to use this table is as follows. When you are writing a complex MATLAB expression, the order in which the parts of the expression should be evaluated is determined by the ordering given in Table 3-6. All operations involving operators at Precedence Level 1 should be evaluated first. All operations at Precedence Level 2 should be performed next, on to finally, all operations at Precedence

Table 3-6: Precedence Rules for MATLAB Operators (scalars only)

Precedence Level	Operator
1	parentheses
2	numerical exponentiation ($\wedge$)
3	numerical unary plus (**+**) numerical unary minus (**−**) logical negation (**~**)
4	numerical multiplication (*****) numerical division (**/**)
5	numerical addition (**+**) numerical subtraction (**−**)
6	relational *all the relational operators* (**<, <=, >, >=, ==, ~=**)
7	logical AND (**&**)
8	logical OR (**|**)

Level 8 should be performed. If two operators are at the same level of precedence, then their evaluation order is a left to right order in the expression. In this table, exponentiation is at Precedence Level 2, multiplication is at Precedence Level 4, and subtraction is at Precedence Level 5. Thus, the evaluation of the complex expression **b^2-4*a*c** gives the desired result.

Let us finish the task of solving for one root of the quadratic relation we set up earlier:

```
>> a = 1;
>> b = 3;
>> c = 2;
>> numerator = -b + ( b^2 - 4*a*c )^0.5;
>> denominator = 2*a;
>> x1 = numerator / denominator
x1 =
          -1.00
```

We could have written this in even more compact form:

```
>> a = 1;
>> b = 3;
>> c = 2;
>> x1 = (-b + (b^2 - 4*a*c)^0.5) / (2*a)
x1 =
          -1.00
```

To understand this last form fully, remember that operators at the same precedence level are evaluated in left to right order. Step through the last examples carefully. The ability to write complex expressions in MATLAB is essential, and that ability depends on your understanding of how to apply the Rules of Precedence.

3-4.3   Synopsis of Scalar Built-In Functions, Operators, and Expressions

Section 3-4 contained a great deal of material. Following is a Section Synopsis for this material to help you keep the information clear and committed to memory. This synopsis is not stand alone but rather should be used as a memory jog for the material.

---

1. **A MATLAB variable can be thought of as a *named container*. The value of a MATLAB variable then is the contents of the box.**

2. **A variable (holding a value) is a centrally important idea in performing some computation because it allows us to decompose a complicated computation to a series of simple computations where each succeeding computation can use the result of the one before.**

3. **Setting the variable value is done using the *assignment operator*; fetching the value of a variable is done by typing the name of the variable.**

4. **Variables come in different flavors (types): scalar numerical variables and scalar logical variables are the two types introduced in this section.**

5. **In MATLAB the type of a variable is defined by the way the variable is used.**

6. **Scalars are simple numbers.**

7. **logic values can be TRUE or FALSE.**

8. **In MATLAB the logic value FALSE is represented by the number 0; the logic value of TRUE is represented by any non-zero number.**

9. **MATLAB operators are a special subclass of MATLAB built-in functions.**

10. **The class of arithmetic operators take numerical variables as input, and output a numerical result. (See Table 3-3 on page 3-51.)**

11. **The class of relational operators take numerical variables as input and output a logical result. (See Table 3-4 on page 3-52.)**

12. **The class of logic operators take logical variables as input and output a logical result. (See Table 3-5 on page 3-55.)**

13. **MATLAB expressions are blueprints for performing computations.**

Synopsis for Section 3-4

3-5   Problem Sets for Scalars

3-5.1   Set A: Nuts and Bolts Problems for Scalars

Problem 3-A.1 (Section 3-1)

Define a variable named **angle** to be 45°. Convert **angle** to radians using the following:

$$radians = degrees \times \left(\frac{\pi}{180}\right)$$

Store the result in the variable **angleRad**. Use the predefined variable **pi** in your calculation.

Using the Command History window, repeat the calculation for **angle** = 60°, 75°, and 90°.

Problem 3-A.2 (Section 3-1)

Suppose we have an independent variable **x** and two dependent variables **y** and **z**. Here are the expressions to calculate **y** and **z**:

$$y = x - 5$$

$$z = 3y^2 - 16y + 74$$

Set **x** to be 5. Calculate the corresponding values of **y** and **z**.

Using the Command History window, repeat the calculations for **x** = 7, 9, and 11.

Problem 3-A.3 (Section 3-1)

The area and the perimeter of a circle of radius **r** are found with the following expressions:

$$circleArea = \pi r^2$$

$$circlePerimeter = 2\pi r$$

Set the radius (**r**) to the value 5. Calculate the area and perimeter of a circle by entering the formulae above.

Now set the radius to be 7. Double-click on line commands in the Command History window to recalculate the area and perimeter.

Set the radius to 9. This time, select both line commands by holding the *Ctrl* key and copy them to the Command Window. Press *Enter* to carry out both calculations.

Problem 3-A.4 (Section 3-2)

What is a scalar? How does a scalar differ from a vector? How does it differ from an array (or a matrix)? Explain in scientific and MATLAB terms.

Problem 3 A.5 (Section 3 2)

Having a magnitude and a direction, force is represented by vectors in science. Suppose we represented a force vector in the Command Window by defining its **x** and **y** components as follows:

```
>> forceX = 3
>> forceY = 5
```

What type of MATLAB variables are **forceX** and **forceY**? Review the definitions of scalars, vectors and arrays in Section 3-2 to verify your answer.

Problem 3-A.6 (Section 3-3)

Using **angleRad** you defined in Problem 3-A.1, find the cosine and sine of **angle**. Store the results in variables **cosAngle** and **sinAngle**, respectively.

Problem 3-A.7 (Section 3-3)

Using variables **cosAngle** and **sinAngle** (Problem 3-A.6), define the following:

$$\alpha = 45°$$
$$z = \cos^2(\alpha) + \sin^2(\alpha)$$

A well-known identity in trigonometry is that the sum of cosine squared and sine squared of an angle always equals one. If your result for **z** was not one, review Section 3-3 and redo the operations.

Problem 3-A.8 (Section 3-3)

Redefine the value of variable **angle** (Problem 3-A.6) as 65° and recalculate the value of **z** (Problem 3-A.7) using the Command History window.

Problem 3-A.9 (Section 3-3)

Create a variable named **z** and set it to 7. Calculate **4z** and store the result in **z**.

Problem 3-A.10 (Section 3-3)

Assuming **x** is a defined variable in the MATLAB workspace, are the following scalar operations valid in MATLAB? If not, why not?

Hint: Remember, '=' is the assignment operator in MATLAB.

Part A   `>> x + 2 = 4`

Part B   `>> x = 4 - 2`

Part C   `>> x = 4 - (-2)`

Part D   `>> 5 = x`

Part E   `>> 5 + 3 = 7 + 1`

Problem 3-A.11 (Section 3-3)

Solve the following problems with paper and pencil. When you are finished, execute the code in the Command Window to verify your answers. Use your understanding of the container model as you think through the questions.

Part A   What do the variables **x** and **y** contain after executing the following lines?

```
>> x = 4

>> y = x

>> x = 6
```

Part B   What does **x** contain after executing the following lines?

```
>> x = 2

>> x^2
```

Part C   What does **x** contain after executing the following lines?

```
>> x = 2

>> x = 2 * x + 5

>> x = (x-3)/3
```

Problem 3-A.12 (Section 3-4)

Set **z** to 5 in the Command Window. Set **k** and **m** using the expressions below.

Part A $k = e^{2z}$

Part B $m = \log(e^z)$

(The mathematical *log* function means "logarithm to the base 10.")

Problem 3-A.13 (Section 3-4)

One way to test for the sign of scalar **x** is to check if $\sqrt{x^2}$ equals x. Assume the following equality holds:

$$x = \sqrt{x^2}$$

If so, then we can conclude that **x** is non-negative; otherwise **x** is negative. (A non-negative number is one that is positive or zero. The relationship above is NOT a MATLAB assignment statement.)

Successively, set **x** to each of the following numbers and use the MATLAB relational equality operator to test whether **x** is non-negative: {-5, 10, 35, -292, 0, 101}.

Problem 3-A.14 (Section 3-4)

One way to check if a scalar is an integer is to round the scalar and compare the result with itself.

The built-in MATLAB function **round** can be used for rounding a number. For instance, **round(5)** and **round(5.3)** will return 5. But **round(5.6)** will return 6.

Using the built-in function **round** and MATLAB relational operators, test if **x** is an integer for the following values of **x**: {-5.2, 10, 35.3, -292, 0, 101.99}.

Problem 3-A.15 (Section 3-4)

(Be sure you have solved Problem 3-A.13 and Problem 3-A.14 before attempting this problem.)

Use relational and logic operators to do the following tests in MATLAB:

Part A	Is 5 a non-negative integer?
Part B	Is 7.7 a non-negative integer?
Part C	Is 7.7 *not* a negative integer?
Part D	Is 5 or 7.7 a negative integer?

Problem 3-A.16 (Section 3-4)

Create expressions using variables **x** and **y** to answer the following questions. So you have explicit values for **x** and **y** to test your work, set **x=7** and **y=5**.

Part A   Is **x** greater than 6 and less than 10? Here is another way to ask the same question: *Is **x** in the range (6,10)?*

Part B   Is **x** greater than or equal to 10, or less than or equal to 6? Here is another way to ask the same question: *Is x not in the range (6, 10)?*

Part C   Is (**x** is greater than 6) *and* (**y** is greater than **x** but less than 10)?

Part D   Is (**x** is in the range (6,10)) but (**y** *not* in the range)?

Part E   Is **x** or **y** in the range (6,10)?

Problem 3-A.17 (Section 3-4)

Suppose that **x=4** and **y=2**. Compute the following expressions in MATLAB.

Part A $3(x-y)$

Part B $\dfrac{x}{2y}$

Part C $\dfrac{y^3}{(1-x^2)}$

Part D $\dfrac{y^3}{(1-x)^2}$

Part E $\left(1-\dfrac{y}{x}\right)^{-y}$

Part F
$$\frac{1}{\dfrac{1}{2y} - \dfrac{y+6}{x-2}}$$

Problem 3-A.18 (Section 3-4)

Suppose that **x=4** and **y=2**. Compute the following expressions in MATLAB:

Part A $\sin\left(y - \dfrac{1}{x}\right)$

Part B $\sin\left(y - \dfrac{1}{x}\right)^2$

Part C $\sin\left(\left(y - \dfrac{1}{x}\right)^2\right)$

Part D $\tan\left(\ln\left(\dfrac{1}{x} + \dfrac{1}{y}\right)\right)$

Part E $e^{-2\sin(x)\cos(y)}$

3-5.2   Set B: Problem Solving with MATLAB Scalars

Problem 3-B.1

A quadratic polynomial such as
$y = 3x^2 + 2x - 6$ defines a parabola in
the two-dimensional space. The vertex
of a parabola is the point at the top if
the curve is concave down or the point
at the bottom if the curve is concave
up.

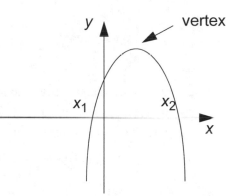

For a quadratic polynomial
$y = ax^2 + bx + c$, the x coordinate of
the vertex and the peak value of the
parabola can be found using this
relationship:

$$x_{\text{vertex}} = \frac{b}{2a}$$

$$y_{\text{vertex}} = ax_{\text{vertex}}^2 + bx_{\text{vertex}} + c$$

In the Command Window, set the values of the constants to be **a=1, b=3**, and
c=2. Calculate the value of x_{vertex}. Calculate y_{vertex}.

Set **b=4**. Recalculate the values of x_{vertex} and y_{vertex} by selecting the
appropriate command lines in the Command History window and carrying them to
the Command Window. Note that you do not need to redefine variables **a** and **c** in
this case since they are already set to their original values.

Set **c=4** and recalculate x_{vertex} and y_{vertex}. For which value of **b** did you calculate
the vertex coordinates, 3 or 4?

Problem 3-B.2

The Newtonian gravitational force between two objects (like planets) in space is calculated using the following formula:

$$F_{gravitational} = -G\left(\frac{m_1 m_2}{d^2}\right)$$

Where G is the universal gravitational constant ($6.672 \times 10^{-11} \text{N} \text{m}^2 \text{kg}^{-2}$), m_1 and m_2 are the masses of the planets in kilograms, d is the distance between planets in meters, and the resultant force is in Newtons.

Using MATLAB, calculate the gravitational forces in the following:

Part A   The Earth ($m_E = 5.97 \times 10^{24} \text{kg}$) and its moon ($m_M = 7.36 \times 10^{22} \text{kg}$), assuming an average distance of 3.844×10^8 meters separating the Earth and its moon.

Part B   The Earth and Sun ($m_S = 1.99 \times 10^{30} \text{kg}$), assuming an average distance of 1.496×10^{11} meters between the Earth and the Sun.

Part C   An Airbus A320 (74 tons) on the ground and the Earth, assuming the radius of the Earth is 6380 km.

Part D   An Airbus A320 at 40,000 feet and the Earth, assuming the radius of the Earth is 6380 km.

Part E   An Airbus A320 at 10,000 miles and the Earth, assuming the radius of the Earth is 6380 km. (Assume the A320 was towed to that location.)

Problem 3-B.3

Money saved in a bank account will compound with the interest according to the following formula:

$$newBalance = oldBalance \times \left(1 + \frac{interestRate}{100}\right)$$

The interest rate is given in percentage.

In the MATLAB Command Window, set **oldBalance** to 1,000 and **interestRate** to 6 percent. Calculate **newBalance**. This is the account balance after one year.

Store the value of **newBalance** in the variable **oldBalance** and drag the command line that calculates **newBalance** to the Command Window. Store the value of **newBalance** in **oldBalance** again. Describe what is now stored in **oldBalance**.

Use your Command History to calculate the balance after eight years. Find the answer in no more than two additional steps.

Problem 3-B.4

Redo Problem 3-B.3. To define balance, use this equation:

$$balance = balance \times \left(1 + \frac{interestRate}{100}\right)$$

Set **balance** to 1,000 and **interestRate** to 6 percent. Using these two variables and your Command History, calculate the account balance after eight years.

Problem 3-B.5

Put values of three and seven in variables (containers) **x** and **y**, respectively. After assigning the values, switch the values of **x** and **y** without using any numbers. Be sure your code for the swap works by checking the results: check the values of **x** and **y** after your swap code is executed.

Hint: You can use an unlimited number of containers in MATLAB.

Problem 3-B.6

The table below shows the prediction of school administration for the percent of students who will be graduating and the number of new admissions for each semester. The enrollment just prior to Fall 2004 is 7,300. What is the expected enrollment in Fall 2006? (Do not worry about fractions of students in your result.)

You should make use of the Command History window when you are doing the same calculation multiple times.

	Number of New Admissions	Predicted Graduation (%)
Fall 2004	1,200	11
Spring 2005	740	19
Summer 2005	145	6
Fall 2005	1,450	12
Spring 2006	810	21
Summer 2006	120	4

Problem 3-B.7

For the circuit setup shown in the figure, the current i (amperes) that flows through R_1 can be found using this expression:

$$i = v\left(\frac{R_2}{R_1 R_2 + R_1 R_3 + R_2 R_3}\right)$$

Where V is the electromotive force (volts) and R_1, R_2 and R_3 are resistances (ohms) as shown in the diagram.

Suppose R_1, R_2, and R_3 are 1, 2, and 4 ohms, respectively. Assume V is 20 volts. Find the current i that flows through R_1.

Pay attention to the precedence rules as you develop the MATLAB scalar expression for the relation above.

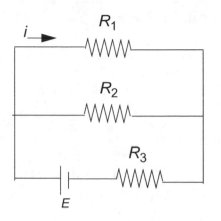

You should do all the tests suggested in the following problems in the Command Window. You may be tempted to investigate the results visually and skip MATLAB tests that use relational and logic operators. However, engineering analysis requires writing code that works on varying inputs, much like how the built-in function **sin** works. You will need to do logical tests involving variables as you progress in MATLAB. You will learn how to write user-defined functions for MATLAB in the next chapter.

Problem 3-B.8

The *geometric mean* (average) of N numbers $x_1, x_2, \ldots, x_N$ is defined by the following equation:

$$geometricMean = \left(\prod_{i=1}^{N} x_i \right)^{1/N}$$

This can also be expressed as:

$$geometricMean = (x_1 \times x_2 \times \ldots \times x_n)^{1/N}$$

Set **x1 = 7**, **x2 = 16**, *and* **x3 = 21**. Find the arithmetic and geometric means of these three numbers. Use MATLAB to test if their arithmetic mean is greater than their geometric mean.

Problem 3-B.9

The roots of a quadratic polynomial, $0 = ax^2 + bx + c$, can be found with this equation:

$$x = \frac{-b \pm \sqrt{b^2 - 4ac}}{2a}$$

Use MATLAB to determine if both roots of the quadratic polynomial $0 = x^2 - 21x + 98$ are greater than 10.

Problem 3-B.10

The unit circle is a useful tool in understanding trigonometric relationships for right triangles. A unit circle is a circle with a radius of one.

For every right triangle whose hypotenuse is 1, the adjacent and opposite side lengths are found with these formulae:

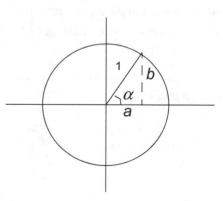

$$a = \cos(\alpha)$$
$$b = \sin(\alpha)$$

Applying the Pythagorean theorem, we find these equations:

$$a^2 + b^2 = 1$$
$$\cos^2(\alpha) + \sin^2(\alpha) = 1$$

Use MATLAB to verify this equality for $\alpha = 0$, 30, 60, and 90 degrees.

The built-in trigonometric functions in MATLAB always take their arguments in radians. You will need to convert the given angles to radians.

Problem 3-B.11

Jim argues that he is taller than Carla and Cindy. But it turns out that Jim is taller than Carla or Cindy, but not both. Jim, Carla, and Cindy are 5.9, 6.1, and 5.7 feet tall, respectively. Use MATLAB to verify the following:

Part A   Jim is not taller than Carla and Cindy.

Part B   Jim is taller than Carla or Cindy.

Part C   Jim is taller than Carla or Cindy but not both.

Problem 3-B.12

Suppose every morning you look out the window and make a decision about how to get to work according to the following rules:

> If it is rainy and the traffic is light, OR if you have fewer than 15 minutes to get to work, you take your car.
>
> If it is rainy and the traffic is heavy, you take the train.
>
> If it is rainy or snowy, you take your umbrella with you.
>
> If it is not rainy and you have more than 15 minutes, you walk.
>
> If it is snowy and the traffic is heavy, you call in sick.

This morning you look outside and see that it is not rainy (**rainyDay=0**) but snowy (**snowyDay=1**), and the traffic is heavy (**heavyTraffic=1**). You have 23 minutes to get to work.

Apply each rule using relational and logic operators of MATLAB and decide if you will take your car (**drive**), take the train (**catchTrain**), walk (**walk**), or call in sick (**callAndCough**), and if you will need to take your umbrella (**takeUmbrella**).

If the result you get does not make logical sense to you, examine the rules again and improve them. Run your example again and see if you get a more sensible result.

Problem 3-B.13

Suppose you are working in your company's accounting department. Your boss gives you a set of rules and asks you to calculate the salary raise for each employee. Here are the rules:

> If an employee has fewer than six years of experience and has evaluations that are positive OR has six or more years of experience and the evaluations are not positive, the employee will get an eight percent raise.

> If an employee has six or more years of experience and has evaluations that are positive, the employee will get a 12 percent raise.

> Otherwise, the employee will not get any raise.

Write two rules in MATLAB and test your rules with the data in the following table. Set **salary=35000**, and apply the MATLAB expressions that implement the company rules.

Employee Name	Years of Experience	Evaluations	Raise (%)
Ms. Moore	5	1	?
Ms. Tenenbaum	11	0	?
Ms. Costanza	3	0	?
Mr. Cane	9	1	?

Now condense your two rules into one that will return the desired percent raise. Test your new rule on the data table again.

Chapter 4

Saving Your Work in MATLAB

You have seen that one way to save your work in MATLAB is to use the Command History window to rerun commands you previously typed to the Command Window (Section 3-3.2 on page 3-44). When working between the Command Window and the Command History window on a given problem, this mode of saving work can be effective. However, if you want to save your MATLAB work across work sessions, or if you are working as part of a group and need to share your work with the group, the *Command History window/Command Window procedure* is not sufficient. MATLAB gives you numerous ways to save your work for retrieval. Advantage can be gained using *encapsulation*, which we will cover in this chapter.

All of the methods of saving work discussed here depend on saving files locally on your computer or on a file server to which you are connected. The first topic we need to cover in this section is how you tell MATLAB where you want to save your file, i.e., in which folder on your local computer or on your file server you want to store your file.

4-1   The MATLAB Current Directory

In Section 3-1, we described parts of the MATLAB environment. In particular, Figure 3-1 showed the main MATLAB window as it first appears. Examine the part of the main window labeled as Workspace or Current Directory. Start MATLAB, and once the main window appears, click on the tab marked "Current Directory." The Current Directory in MATLAB is the folder on your computer (or a connected file server) that MATLAB uses for all operations that save files unless you tell MATLAB to use a different folder. The main window of the MATLAB environment is shown in Figure 4-1 with the major tools for using the current directory marked.

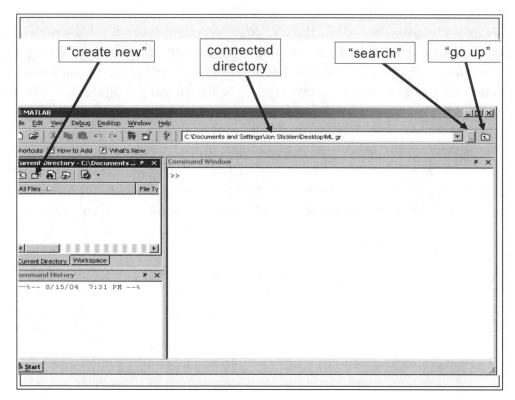

Figure 4-1: Current Directory Window (note operations labels)

In Figure 4-1, the box labeled "connected directory" is a display area showing the folder that MATLAB is using as the default folder for saving and locating files. The button marked with an ellipsis (and labeled "Search") brings up a standard Windows (or Mac or Unix) file locator browser. Use this button when you want to locate a folder on your machine (or a connected file server) that you want to target as the new connected directory.

The folder icon with an up-arrow on it (labeled "go up" in the Figure 4-1) moves up one level in the computer's folder structure. The folder icon labeled "create new" in Figure 4-1 creates a new folder inside the Current Directory.

Commands are in MATLAB that may be used for all of these interactive features. Look at the command that follows:

```
>> cd C:\Documents and Settings\Jon Sticklen\My Documents\ML
```

This resets the Current Directory to be a folder called ML in My Documents on my personal PC, if it exists (note that the path to "My Documents" folder on youur PC will most probably be different). Through most of what you will do in this course, the interactive features of the MATLAB interface will suffice.

Open **My Documents** on your computer. Create a new folder in **My Documents** (or someplace of your choosing) that is called **ML_Work**. Make sure you have MATLAB running. Use the ellipsis button to find your newly created **ML_Work** and make it the Current Directory. Because you have not created any files in **ML_Work** yet, the window will show no files. Your Connected Directory window should look like that shown in Figure 4-2.

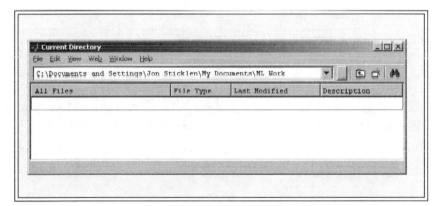

Figure 4-2: Empty Current Directory Window—The Starting Point

We are showing the "white box" that displays the name of the current directory with the current directory window pane so we don't have to show the entire MATLAB window containing all panes. We will do this throughout the discussion in this chapter.

1. **The Current Directory is the default folder to which MATLAB will save all files.**

2. **The Current Directory is the folder that MATLAB will search first when it wants to load any files.**

3. **The Current Directory may be changed interactively using the Current Directory window, as well as by use of built-in MATLAB commands.**

Synopsis for Section 4-1

4-2 Saving MATLAB Commands in Script Files

Saving work as **script files** will be the first method we examine for saving your work in MATLAB. Script files are simply a set of MATLAB commands bundled together into one file. Executing a script file has the same result as executing each and every command stored in the script file in the MATLAB Command Window.

Bring up the MATLAB editor by executing the following command:

```
>> edit
```

Select **Save** under the File menu of the editor window. Use the dialog box to save a new empty file (you have put nothing into the editor yet) called **MyFirstMATLABFile.m**. The extension (**.m**) is important. **DOT-M** files are MATLAB code files (or source files). Your Current Directory window should look like the one in Figure 4-3.

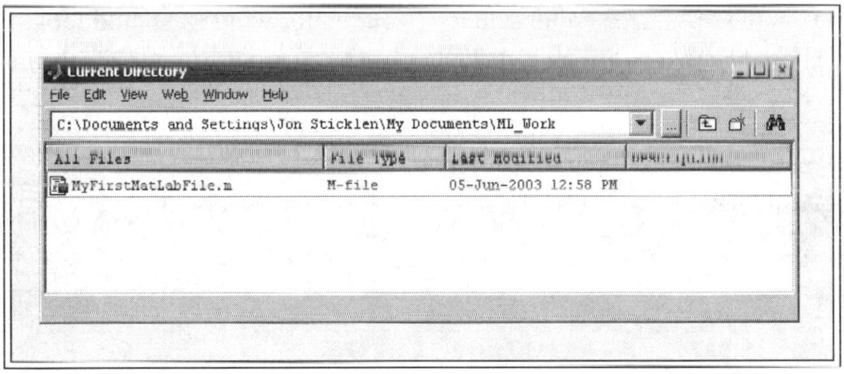

Figure 4-3: Current Directory After Saving One File

Suppose our problem is to compute the volume of a hemisphere given its radius and the surface area of the same hemisphere. The commands you might execute in the Command Window for solving the problem for a hemisphere of four-feet radius could be as follows:

```
>> rad               = 4;                  % feet
>> volHemisphere     = (2/3) * pi * rad^3;% cu ft
>> surfAreaHemisphere = 2 * pi * rad^2;  % sq ft
>> rad, volHemisphere, surfAreaHemisphere
rad =
          4.00
```

```
volHemisphere =
          134.04


surfAreaHemisphere =
          100.53
```

We wrote the first three commands to the Command Window with semicolons so no answer would be returned from those lines, wrote the final command by naming the variables we wanted to see the values for, and separated them by commas so we could put all three instructions in the same command line.

How do we take the four commands from above, put them into a file, and save the file for later execution as a single piece of code? Copy the commands and put them into the empty file we created: **MyFirstMATLABFile**. Do not copy the ">>" to the editor window; they are part of the Command Window only. After typing these four commands in the MATLAB editor window, your editor window should look approximately like that in Figure 4-4. The only additions to the code are the comment

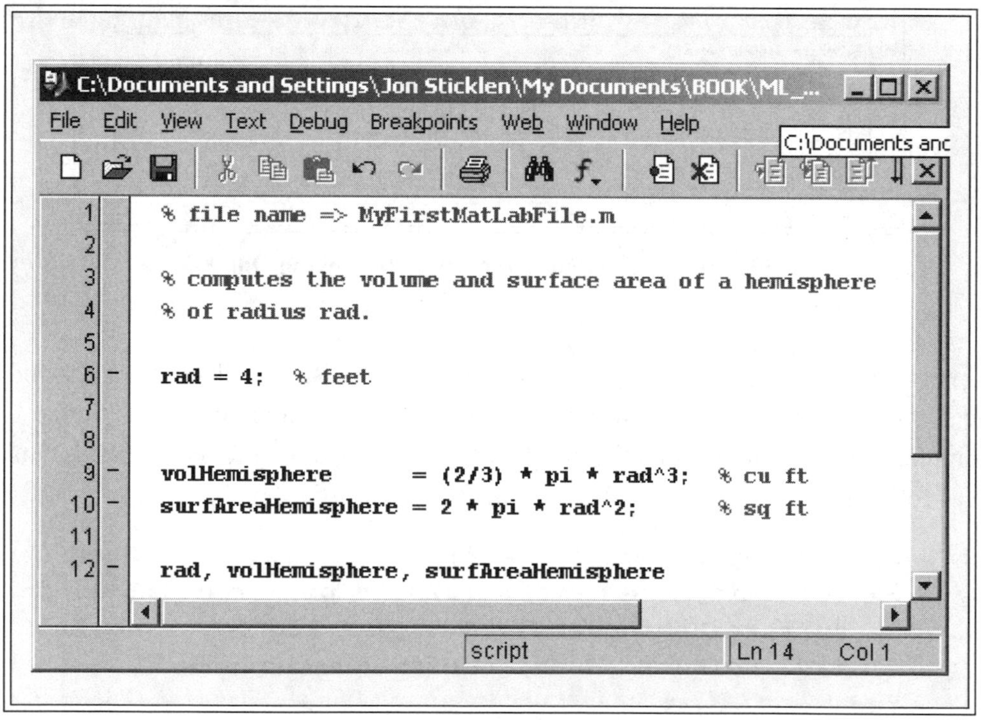

Figure 4-4: MyFirstMATLABFile

lines at the top of the editor window. These comments, which you should take as standard for a script file, are the file name and a brief description of what the file accomplishes. Having constructed the script file, choose **Save** under the **File** menu of the editor window to store your file. Close the editor window.

In the MATLAB Command Window, execute the following command: **MyFirstMATLABFile**. MATLAB will display any results generated from executing your script file.

```
>> MyFirstMATLABFile

rad =
        4.00

volHemisphere =
        134.04

surfAreaHemisphere =
        100.53
```

When you execute **MyFirstMATLABFile** as a command in MATLAB, MATLAB tries to find a file in the Current Directory named **MyFirstMatlabFile.m**. If the file is found, then MATLAB executes each line of code in the file as though it were typed into the Command Window. Thus, a script file in MATLAB is a way of packaging a number of commands. The packaging (into a *DOT-M* file) allows running all the lines of code as a block. This idea of a code block and thinking of it as a single package is important, and it will come up often. Translated into computer science terminology, the word that describes this packaging is *modularization*. We turned what were three commands into one command that can be called from the Command Window in **MyFirstMATLABFile**. Stated another way, we created a single module.

The point? There are several:

1. We have stored the code in a permanent place (the file) so we can reuse it later if needed.

2. Since the code is in a file, we can transfer the problem solution (the file containing the code) by moving the file.

3. We have turned the solution we set into a one-line command that can be executed from the Command Window. This last point is the point about modularization.

Suppose we wanted to ask the following question:

> *What value for the radius of a hemisphere (in meters) will produce a volume
> of the hemisphere (in cubic meters) that is numerically equal to the surface
> area of the hemisphere (in square meters)?*

We can modify **MyFirstMATLABFile** by removing line six so the radius of the
hemisphere is not set in the script file, but rather set in the Command Window.
Requiring that the radius of the hemisphere be set outside the script file should be
recorded in the script file comments, so the modified script file should look like that
shown in Figure 4-5. Open **MyFirstMATLABFile** in the editor window. Modify the
commands in the script file so it is like the set of commands in Figure 4-5. We want to
name this file differently: **MySecondMATLABFile**. After you have completed the
modifications to the script file, pull down the **File** menu, select **Save As**, and give
MySecondMATLABFile.m as the file name. Your Current Directory window should
now show two files.

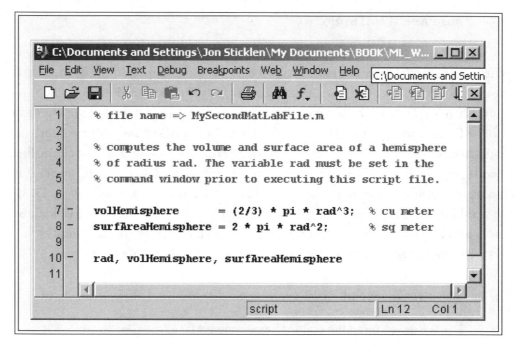

Figure 4-5: Modification of MyFirstMATLABFile

Type the following commands and execute:

```
>> rad = 4; % meters
>> MySecondMATLABFile
```

The result you will see in your Command Window is as follows:

```
rad =
            4.00
volHemisphere =
         134.04
surfAreaHemisphere =
         100.53
```

The second version of the script file is similar to the first. The only computational difference is that the variable **rad** is set outside the script file. In a sense, **rad** is handed to the script file for calculation. One important difference is to a person reading the file and using it. An additional comment indicates that before the script file can be used, the variable **rad** must be set in the Command Window. The script file **MySecondMATLABFile.m** will not work properly if **rad** is not set.

The guess of the value for **rad** of four did not produce the answer for which we are searching. Suppose we try a value for **rad** of two (meters).

```
>> rad = 2; MySecondMATLABFile

rad =
            2.00

volHemisphere =
          16.76

surfAreaHemisphere =
          25.13
```

We collapsed what is typed by putting two commands on the same line. When more than one command is put on a single command line, then the commands must be separated by a comma or a semicolon. If a comma is used, then the result of the command is displayed in the Command Window. If the semicolon is used, no output is generated to the Command Window. (Why does the Command Window show a listing of the value of **rad**?)

But, we do not have the value of **rad** that gives the desired result: The volume of the hemisphere and the surface area of the hemisphere are numerically the same. We have learned something though. When we tried **rad = 4**, the value of the surface area was smaller than the volume. When we tried **rad = 2**, the value of the surface area was larger than the volume. What does this tell us? Why? Try other values for **rad** to see if you can solve the problem.

An easy algebraic solution to the problem exists; find the solution using algebra, and use paper and pencil as a check on your MATLAB script. (Hint: You have two equations, one for the volume of the hemisphere as a function of its radius and one for the surface area of the hemisphere as a function of its radius. You want the values to be numerically equal. Set the right-hand sides of the two equations equal, and solve for the radius.)

Here is a slightly more difficult problem with which we will end this section:

> *Compute the cost of a water tower. The water tower is made of two major parts: a "tower" that is a right circular cylinder and a "cap" that is a hemi- sphere. The tower cross section has a diameter of* **diam** *meters; the height of the tower section is* **hght** *meters. The cost (materials and labor) for the water tower is $300 per square meter for the tower section and $500 per square meter for the cap section. Compute the volume of the completed water tower. Assume* **diam** *is 10 meters, and* **hght** *is 30 meters.*

The first task is to make a sketch of the water tower and label all relevant dimensions. This *problem visualization* is an important step for this example and, in general, is a step you should take seriously for most technical problems. Problem sketches need not take a long time or look professional. Of course, if such a sketch is to become part of a formal report (e.g., to your boss), then it should look professional and probably be done on a computer. But if you are using a problem sketch to solidify your own thinking about a problem, then a rough paper and pencil sketch is fine. For our water tower problem, a sketch like the one shown in Figure 4-6 hits the mark because it conceptually shows the problem.

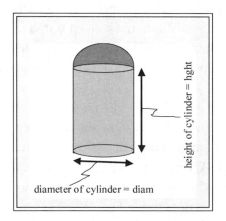

Figure 4-6: Sketch of Water Tower

Our prescription was to make the sketch and label all the relevant dimensions. In the drawing of Figure 4-6 we marked two of the dimensions of the problems: the height of the cylinder and the diameter of the cylinder. How about the diameter of the hemispherical cap? If we are going to compute the volume and surface area of the hemisphere, we are going to need the size of the cap (its diameter), too. But the sketch makes the situation obvious: for a water tower to be composed of a cylindrical tower and a hemispherical cap, the cap and the tower must have the same diameter. A sketch, even for a simple problem, can make a connection that in retrospect is obvious, but may not be obvious before you make the sketch. An initial sketch should include only the problem dimensions which are given in the problem statement. A second sketch (that might be an addition to the first) can indicate other problem dimensions that must be computed from those given in the problem statement. You should make it clear what dimensions are given in the problem statement, as contrasted to the dimensions you will have to compute. The sketch in Figure 4-7 suffices.

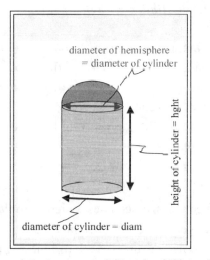

Figure 4-7: Augmented Sketch of Water Tower

Having done the sketch, the next step is to lay out the relevant background material you are going to need. For this problem, that material consists of equations for the cylinder's surface area and volume and the hemisphere's surface area and volume. The next step is to assemble the background knowledge and lay it out. Sometimes you will know the material you need to solve a problem. Many times you will not remember it all or have possibly studied all required parts. This is part of what an engineer deals with on a regular basis: the need to learn more than he or she currently knows or can remember to solve the problem. For your purposes, you can find most background material by using the Internet. Google is a good search engine to find what you need. Use Google now to find information about right circular cylinders and hemispheres.

For the cylinder, you need the volume (given the diameter and height) and the surface area (given the diameter and height). For the hemisphere, you need the volume (given the diameter) and the surface area (given the diameter). The relations you could find on Google are as follows: Equation 4-1 for the hemisphere and Equation 4-2 for the cylinder.

$$volumeHemisphere = \frac{2}{3}\pi \left(\frac{diam}{2}\right)^3$$

$$surfaceHemisphere = 2\pi \left(\frac{diam}{2}\right)^2$$

(EQ 4-1)

$$volumeCylinder = (hght)\pi \left(\frac{diam}{2}\right)^2$$

$$surfaceCylinder = (hght)\pi (diam)$$

(EQ 4-2)

Next, convert these mathematical relations of Equation 4-1 and Equation 4-2 into MATLAB commands and package them in script files, and package the cost factors and calculations to compute the cost and volume of the water tower. The next issue is how to break up the problem into chunks, where some of the chunks are going to become the script files we will build. For this problem, we could write one script file that would do the entire job for us.

So, why worry about breaking the problem into pieces? The reason is that by breaking a problem into manageable pieces, it becomes easier to implement in MATLAB and to debug written code. Thirty years of experience in a branch of computer science called *software engineering* has shown me that the job of building programs in MATLAB or any other computer language becomes easier if the job is broken down, or in computer science lingo, if the problem is decomposed.

Suppose we write how we intend to solve the water tower problem:

1. Calculate the surface area and volume of the hemispherical cap of the water tower.
2. Calculate the surface area and volume of the cylindrical part of the water tower.
3. Add together the volumes to get the total volume of the water tower.
4. Use the surface areas to calculate the cost of the water tower.

Steps 1 and 2 of this decomposition deal with different solid geometry objects: Step 1 with hemispheres and step 2 with cylinders. Because the two steps deal with distinctly different physical object types, you should implement each step in a separate script file. We have decomposed the first two steps of our problem in two script files shown in Figure 4-8.

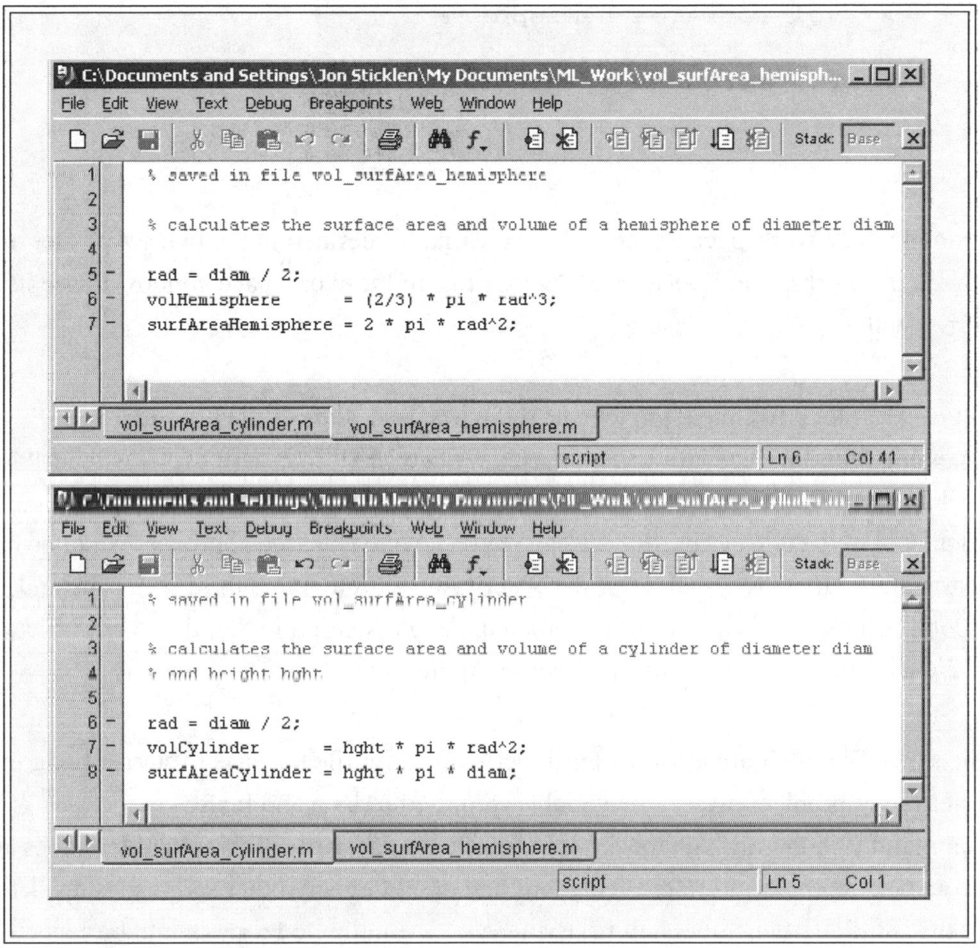

Figure 4-8: Script Files for Parts One and Two of the Water Tower Problem

Using your MATLAB editor window, create and save the two script files as shown in Figure 4-8. In your MATLAB Command Window, execute the following:

```
>> clear, clc
>> diam = 10;  hght = 30;  % both in meters
```

clear removes all defined variables from the workspace, and **clc** clears the Command Window. Before starting a new problem, you should clear the workspace to be sure you do not mix variables you used for earlier problems with your current problem.

Examine your Command History window. You should see two variables: **diam** and **hght**. Execute the following commands:

```
>> vol_surfArea_hemisphere

>> vol_surfArea_cylinder
```

Examine your Workspace window. All the variables defined in the two script files are now in the workspace. Look at **volCylinder** in the Workspace window to see its current value.

Let us take stock. We have implemented the first two steps of the problem decomposition for the water tower problem in two MATLAB script files. We need to implement the last two steps of the decomposition. These two steps use values calculated in the two script files we have built, so a reasonable approach would be to tie together our entire solution for the water tower problem in a third script file and to accomplish the two last steps of the problem decomposition in that third script file. One cut on the third script file is shown in Figure 4-9.

The script file in Figure 4-9 is called a top-level script file because to start solving the water tower problem, we set the variables required (**diam** and **hght**) in the Command Window and run the top-level script. Notice, **waterTowerProblem** does all the work of both performing calculations and of calling any other script files needed for the solution. Remember before starting a new solution, **clear** should be executed to clear all old variables. Implement **waterTowerProblem** as a script on your computer and save it. Then execute the following:

```
>> clear

>> diam = 10; hght = 30;

>> waterTowerProblem
```

After execution, MATLAB will respond with the answer to the water tower problem:

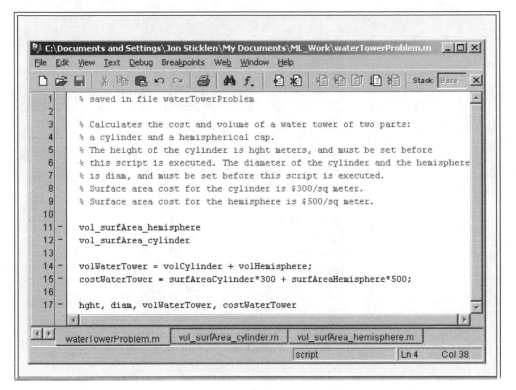

Figure 4-9: Top-Level Script for Solution to Water Tower Problem

hght =

   30.00

diam =

   10.00

volWaterTower =

  2617.99

costWaterTower =

  361283.16

The results showing in your Command Window are from Line 17 in the script file **waterTowerProblem.m**. (See Figure 4-9.)

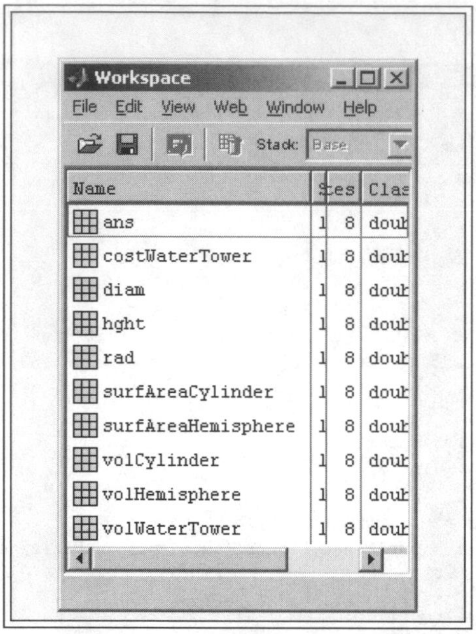

**Figure 4-10: Workspace After Executing
Script: WaterTowerProblem**

Examine your Workspace window. It should look like that shown in Figure 4-10.
Every variable you created in the three script files are in your workspace. In fact, we
relied on this point in our solution: All variables created in a script file are available
once the variable is defined. The **waterTowerProblem** script depends on the
availability of variables defined in the **vol_surfArea_cylinder** script and the
vol_surfArea_hemisphere script.

But there is a problem lurking. When we wrote the **waterTowerProblem** script, we
had to know the names of the variables calculated in the **vol_surfArea_cylinder** and
the **vol_surfArea_hemisphere** scripts. Suppose that the problem were a big problem.
For example, consider the task of building the code for Microsoft Windows, a task
involving tens of millions of lines of code. Each programmer would have to work
independently on his or her piece of the problem. Once complete, all the pieces would
be put together. If that process of composition requires that we know the names used
internally for the variables in each script file, we would have a difficult problem on
our hands indeed. We will see a way around this problem in the next section on
MATLAB's user-defined functions.

1. A script file consists of groups of MATLAB commands bundled together into a *module*.

2. A script file must have a *DOT-M* extension.

3. A script file is created in the MATLAB editor and saved to the current directory.

4. When executing inside a script file, all variables in the workspace are available.

5. All variables created in a script file are available in the workspace at the Command Window or in other script files.

6. Making a sketch of a problem is important in making concrete the problem context.

Synopsis for Section 4-2

4-3 Saving MATLAB Commands in User-Defined Function Files

In the last section, you learned how you can save a group of MATLAB commands in a script file and how to run all the commands in a script file by typing the name of the script file in the MATLAB Command Window. In this section, you will learn another way to save a set of MATLAB commands, which is saving commands in a *user-defined function* file.

In Section 3-4.1, you encountered one type of MATLAB function: built-in functions. The user-defined functions in MATLAB are the same as built-in functions with one exception: user-defined functions are created by the MATLAB user, i.e., by you. When you start up MATLAB, it comes with a large number of functions for you to use. But you have the ability to extend MATLAB by creating new functions of your own. For the remainder of this section, we will refer to "function" to mean user-defined function even though most of what follows in this section applies equally well to built-in functions and user-defined functions.

The idea of a function is similar to the idea of a script. You use the same MATLAB editor window to create a function file. MATLAB function files are all *DOT-M* files. You tell MATLAB to run a function file the same way you do for a script file by typing the function file name.

On the other hand, script files and function files differ in ways that you as a MATLAB program builder have to deal with. The root difference has to do with when variables can be "seen." All variables you define in a script file are available for use in the Command Window through the common workspace.[1] An example of an assignment statement in a script file named **MyScriptFile** is the following:

```
myVariable = 333 * pi;
```

It creates the variable container called **myVariable** (if it does not already exist) and puts the value 333π into the container. After **MyScriptFile** is run, that named container (**myVariable**) and the value it holds are available in the workspace and in any other script file. The flip side of this is true: A variable defined by executing an assignment statement directly in the Command Window can be seen (which means its container can be seen, and its value accessed and reset) in another script file.

For any function, you must remember three central points:

---

1. Remember, you define or create a variable in MATLAB by assigning the variable a value.

1. The only way of getting a variable's value into a function is for that variable to be input to the function. The variables whose values are "handed to" a function are called *input variables*.[2]

2. The only way of getting a value out of a function is for that variable to be output from the function. Variables whose values are "handed back" from a function are called *output variables*.

3. A variable used inside a function that is not an input variable or an output variable is not visible outside the function. These variables are called *local variables*.

When we described MATLAB script files in the previous section, we emphasized that the Command Window and all script files access variables through the workspace. We have to extend the concept of that workspace to handle functions. Remember, the workspace is a group of variables, named containers, with each named container holding a value.

We need to understand that *there is more than one workspace in MATLAB*. The workspace we have been dealing with for all script files is more properly called the *Global Workspace*. This is the workspace that is used by the Command Window and all script files.

When any function is run, it sets up its own private workspace that holds variable containers that can be seen only by it and only when it is running.

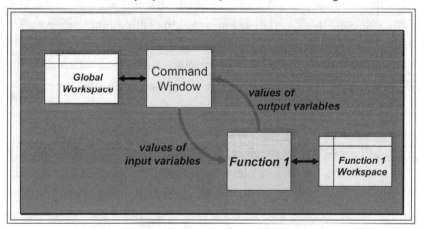

Figure 4-11: Each Function Has its Own Private Workspace

---

2. Another way is by defining a variable to be *global*. Global variables can be assigned in one function, and the values of the variable used in another function. In earlier computer languages, notably early versions of FORTRAN, heavy use of global variables was made for a number of technical reasons. Experience in software engineering has shown that the use of global variables should be selective and restrictive. In this book, we will not introduce the use of global variables.

What we have said so far about functions is summarized in Figure 4-11. Access to a workspace (getting the value inside a variable container or setting the value of a container) is shown by the arrow leading into *Function 1*. When you use a variable in the MATLAB Command Window, the global workspace is accessed; when you access a variable while *Function 1* is running, you access the workspace that belongs to *Function 1*. Variable values for input variables are handed from the Command Window to *Function 1* when MATLAB is told to execute *Function 1*. And after *Function 1* executes, the values for output variables are handed back from *Function 1* to the Command Window.

In MATLAB, variables (or variable names) are not handed from the Command Window to a function (input variables) or handed back from a function after it has run (output variables). Rather, MATLAB passes the values of variables to functions to be executed.[3] But that leaves us with a problem: When *Function 1* (in Figure 4-11) is run, and the Command Window passes it the input variable values, then what should those values inside *Function 1* be called? In other words, what named containers in the workspace of *Function 1* should contain which input variables? A variable value, like a simple scalar number, cannot be manipulated in a function unless it is contained in a named variable. The problem did not exist when we were dealing with script files because variables in use in a script file are all in the global workspace. To understand functions, we must deal with this problem.

The solution lies in making a link between input variables at the Command Window level and the values of those variables that are at the called function level. To see how this linkage is made, we have to start with the way a function is called for execution. In the Command Window, you could call the built-in function for sine, on the value $\pi/2$, and set the value obtained to be the value of the variable **myVar**:

```
>> myVar = sin(pi/2);
```

The way you call user-defined functions is the same. You write the name of the function, followed by a list of variables whose value you wish to pass to the function. The list of input variable names is set off inside a parenthesis pair, and if there is more than one input variable, the input variables are separated by commas. Suppose we have a defined a function called **myAverage** that takes three scalar variables as input, computes the average of the three scalars, and outputs the average. (You'll see

---

3. This type of communication of variable values is called "pass by value." This is the only mechanism MATLAB offers for communication of variables between a caller and a called function. Some programming environments offer other ways of passing variables, including "pass by reference" in which the names of the variables are passed rather than the values.

how to define this function in detail shortly.) From the Command Window, you could set up the three variables you want to average, and call **myAverage**:

```
>> num1 = 3; num2 = 7; num3 = 6;

>> aveThree = myAverage(num1, num2, num3);
```

The variables **num1**, **num2**, and **num3** arc sct up as named containers in the global workspace, and this is to the crux of the issue. How do we hand **myAverage** the values of **num1**, **num2**, and **num3** so the numbers can be averaged in **myAverage**?

In Figure 4-12, a definition (the *DOT-M* file) for **myAverage** is shown. Line 1 in the *DOT-M* definition has a lot of similarities with the second line of the code fragment above where we called **myAverage** from the Command Window.

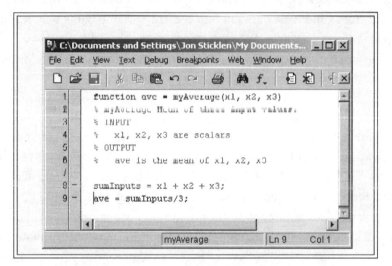

Figure 4-12: Function File for myAverage

In Figure 4-12, after the first word in Line 1 (**function**), the variable name (**ave**) is followed by the sign for the assignment operator (**=**), followed by the name of the function, and followed by an input argument list. The names of the variables used in the Command Window that call to **myAverage** and the names used in the function itsclf are different.

The answer to the situation is that the arguments that are used in the Command Window are linked with arguments in the definition of **myAverage**. When **myAverage** is called from the Command Window, a link is made between the *actual parameters* in the call and the *formal parameters* in the definition of **myAverage**. The linkages are shown in Table 4-1.

**Table 4-1: Argument Linkage Between
Command Window and Call to myAverage**

in the Command Window	variable type	in myAverage
num1	input =>	x1
num2	input =>	x2
num3	input =>	x3
ave3	<= output	ave

When **myAverage** is called from the Command Window, the value of **num1** is taken from the global workspace and a variable container called **x1** is created in the workspace of **myAverage** with its value set to the value taken from **num1**. Similarly, the values for **num2** and **num3** are put into created variable containers **x2** and **x3** in the workspace of **myAverage**. When **myAverage** completes, after **ave** is computed, the value of the container is returned to the Command Window, and this value is put in the **ave3** container in the global workspace.

If you have not implemented **myAverage** yet, do it now. Create **myAverage** as shown in Figure 4-12. Left-click to the right of the line number "8" in your MATLAB editor window. You should see a small red circle appear to the right of "8" that looks like that shown in Figure 4-13. What you did by setting that red circle was to tell MATLAB to Stop before the command on line eight is executed. MATLAB gives you this to debug script files and function files. Later in this chapter, we will cover MATLAB's debugging tools more completely. For now, the goal is to interrupt MATLAB in mid-execution of **myAverage** so we can examine the workspace when MATLAB gets to line eight.

Type and execute the following into your MATLAB Command Window:

```
>> clear, clc
>> num1=22; num2=57; num3=69;
```

Examine your Workspace window. You will see variables **num1**, **num2**, and **num3** defined. Execute the following:

```
>> myAve = myAverage(num1, num2, num3)
```

You will be examining a number of your MATLAB interface windows in turn. Your workspace, command, and editor windows should look like those shown in Figure 4-14. In your Workspace window, the variables that were there before for **num1**, **num2**, and, **num3** are no longer visible. You will instead see variables named **x1**, **x2**, and

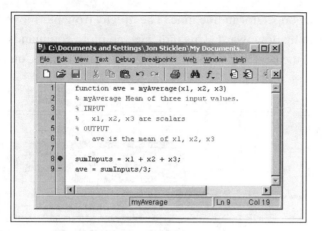

Figure 4-13: **myAverage** with Stop Set

x3. Double-click on the variable **x1**. You will find its value to be 22. Likewise, by double-clicking the variables for **x2** and **x3**, you find that **x2** has a value of 57 and **x3** has a value of 69. This illustrates the linkage that MATLAB makes for you when you run a function. The single Workspace window of the MATLAB interface shows the workspace that is current. When MATLAB was executing a command from the Command Window, the workspace being shown in the Workspace window was the global workspace. When MATLAB is executing a function and the function execution is put on hold, the workspace showing in the Workspace window is the workspace owned by the function.

Look at your Command Window. The normal prompt that MATLAB gives you (>>) has been replaced with a different prompt (**K>>**). The changed prompt (**K>>**) indicates that MATLAB has been interrupted and is awaiting your commands.

Look at your editor window. Next to line number eight, you will see the small red circle that you placed there earlier. You will see a green right-pointing arrow that points to the code of line eight. This indicates that MATLAB is holding before executing line eight and is awaiting commands from you.

In your editor window, pull down the menu labeled Debug and select Step. (The keyboard equivalent of Step is function key F10 in a Windows system.) This has the effect of executing one MATLAB command, which is the command at line eight. In the workspace of the **myAverage** function, a new variable has been created whose name is **sumInputs**. The green arrow (indicating where MATLAB is running in **myAverage**) has moved to line nine. Step ahead again. Another new variable is created in the workspace of **myAverage**, and this one, named **ave. ave**, was created from the assignment operator in line nine. The green arrow has moved forward, but this time it is pointing downward. This indicates no more commands are to be run in **myAverage**, and the next step will take MATLAB back to the Command Window.

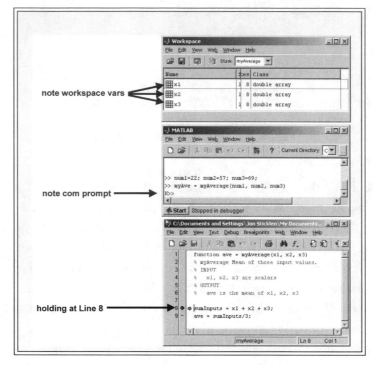

Figure 4-14: Execution Holding at Line 8

Step ahead again. The Workspace window shows the global workspace, and the normal prompt in the Command Window (**>>**) has been restored. The command you gave MATLAB to begin with to start the **myAverage** function was the following:

```
>> myAve = myAverage(num1, num2, num3)
```

The result of executing this command appears in your Command Window:

```
myAve =

      49.33
```

The variable named **myAve** appears in the Workspace window, which shows the global workspace.

We used a small part of the MATLAB debugging machinery here just enough to allow us to interrupt MATLAB while it was running **myAverage** so we could see the

workspace owned by **myAverage**. We will describe more MATLAB tools for
debugging scripts and functions later in this chapter.

The fuss over this simple averaging example (that you could work out in your head) is
not about the specific example, but rather about the idea of functions owning their
own, private workspace. Functions are insulated from the outside world. The only way
into a function is through input variable values, and the only way out is from output
variable value(s). Any other variable used in a MATLAB function can be seen inside
the function, that is, while it is running.

Now you are ready for the nuts and bolts of how to define a function and how you
make the linkage between variables "on the outside" and variables "inside" a function.
Conceptually, the whole story is shown by the example back in Table 4-1. Remember,
variables are not being linked, but are passing from outside of a variable value to
inside where that value is put inside a variable container known in the function.

Let us examine the command line that called **myAverage** and the first line in the
myAverage function file: Figure 4-15.

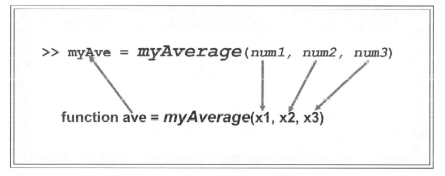

**Figure 4-15: Linkages Between Actual Parameters
and Formal Parameters**

The input-output variables in the command line that calls the function **myAverage**
are called *actual parameters*. The input-output variables in the function definition (in
the function file) are called *formal parameters*.

You should know two things about the linkages between the actual parameters and the
formal parameters. *First*, the number of input variables in the actual parameters is the
same as the number of input variables in the formal parameters, and the number of
output variables in the actual parameters is the same as the number of output variables
in the formal parameters. Simply put, the function expects a certain number of input
variables; if you want to call the function, you have to give the right number of inputs.

The same is true for output though, since we are using one output variable, this is not an issue.

Second, the order of input variables must be the same in the formal parameters list and in the actual parameters list. In **myAverage**, our goal was to find the arithmetic average (the mean) of three scalars. The order of the inputs made no difference, i.e., the average of 3, 4, and 5 is the same as the average of 5, 4, and 3. However, **myAverage** is a special case.

To illustrate the more general case (when the order of arguments makes a difference), suppose we want a function for computing the projection of a vector $\vec{r}$ along the *x*-axis. $\vec{r}$ has a length $|\vec{r}|$, and makes an angle of ϕ with the *x*-axis.

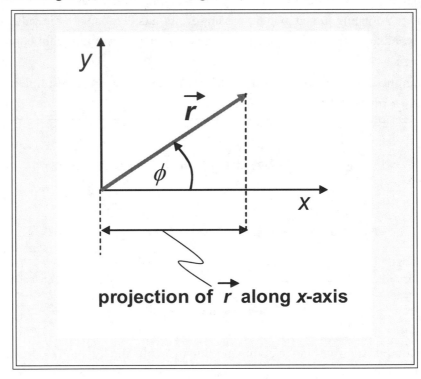

Figure 4-16: Projection of Vector Along the *x*-axis

The situation is shown in Figure 4-16. The relation we want to calculate is $r_{alongx} = |\vec{r}|\cos(\phi)$. We will call the function to perform this operation **projectVectorAlongX**; it is shown in Figure 4-17. A call from the Command Window to run **projectVectorAlongX** might be called the following:

```
>> lenV = 22;     % meters

>> ang = 30;      % degrees

>> projZ_X = projectVectorAlongX(lenV, ang)
```

We have added a little more interest to **projectVectorAlongX** by specifying the angle handed to the function is in degrees; the angle expected by all built-in MATLAB trig functions is in radians.

In **projectVectorAlongX**, the order of the input arguments handed to the function is definitely important. **projectVectorAlongX** expects the first actual input argument to be the value of the vector length, and the second actual argument to be the value of the angle the vector makes with the *x*-axis. If when you call **projectVectorAlongX**, you mix the order of the arguments, you will get an incorrect result.

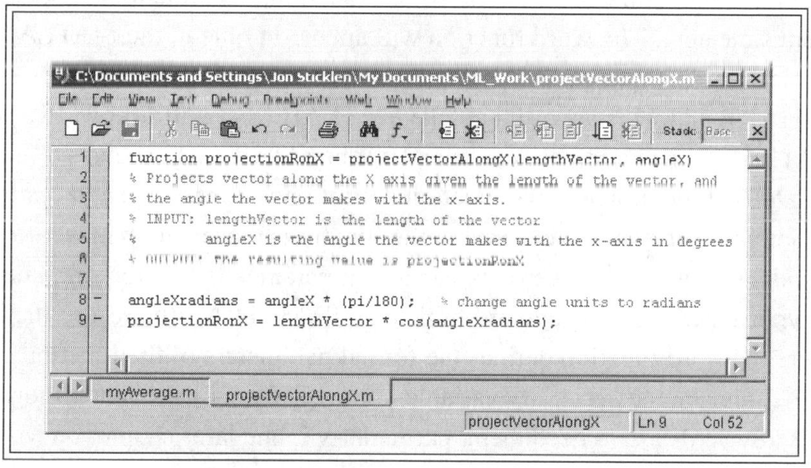

Figure 4-17: Function projectVectorAlongX

Try the function by implementing **projectVectorAlongX**. Call it from the Command Window using these commands:

```
>> lenVec = 32; ang = 60;
>> aProjection = projectVectorAlongX(lenVec, ang)
```

This is the answer you should get:

```
aProjection =

      16.00
```

Try the following:

```
>> aProjection = projectVectorAlongX(ang, lenVec)
```

The answer the Command Window will hand you is 50.88. This answer is incorrect. The order of the input variables in the calling command must be the same order as the input variables in the function file.

So, two factors must match up between the call to a function and the function definition:

1. The number of actual parameters (input and output) must be equal to the number of formal parameters in the function definition.

2. The order of the actual parameters must match the order of the formal parameters in the function definition.

You need to know more details about functions. Look back at Figure 4-15. In the first line of the function definition, after the word "function," the line looks like a regular assignment statement. The word function will appear in blue in the MATLAB editor window and is an example of a MATLAB *keyword*. A keyword (which appears in blue in the MATLAB editor) is a word that tells MATLAB what will follow. In a way, it helps MATLAB understand your commands. When MATLAB executes a function, it loads the *DOT-M* file in which the function definition is stored. Then, it reads the MATLAB commands you have stored in the file, turns those commands into machine runable code, makes the linkages between actual parameters and formal parameters, and runs your function. The keyword function tells MATLAB the rest of the line of code where the word function defines the formal parameters of the function so MATLAB can make the necessary variable linkages. We will encounter more MATLAB keywords as we need them, particularly in the later chapter on MATLAB conditional and iterative constructs.

Second, and you should think of this as a rule you must live with, the name of a MATLAB function must match the name of the *DOT-M* file that contains the definition for the function. If you want to create a function named **jonsMLfunction**, go to the editor window, write the function definition line (as line 1) of the function[4], write in comments that explain what the function does and what its input and output arguments are, and write the MATLAB commands that allow

---

4. You could write lines of comments before the line that begins with function. You may not however write any MATLAB commands before the function line. For stylistic reasons, we will take the position that the function line must be line one in a function file.

the function to do its computational job. You must save the function in a file called **jonsMLfunction.m**. Why is this such a hard and fast rule? Go back to the previous paragraph and see if you can figure out the reason before you read the next paragraph.

The reason the function definition file must be named to match the function name is embedded in the way MATLAB processes functions. When you call a function, MATLAB has to find the file that holds the function definition. By requiring the function file name match the function name, MATLAB can perform this first task easily. If you don't have consistent names between the function file and the function name, MATLAB will not be able to run your function because it will not find the right function file. You must keep the function name and the name of the function file consistent when you write functions in MATLAB.[5]

So far, we have examined functions that have one output variable. We are not limited however in MATLAB to one output variable. As an example, suppose we wanted to modify the function shown in Figure 4-17 so it would output the projection of the vector along the x-axis and the projection of the vector along the y-axis. The modified function **projectVectorXY**, shown in Figure 4-18, will do the trick. In the function line of the modified function, two output variables exist and are separated by a comma; the output variable list is enclosed by square brackets.

If we want to call the modified function, and have its output directed to variables **x** and **y** in the global workspace, we will have to execute the following command:

```
>> [x,y] = projectVectorXY(32, 60)
```

Here is the result:

```
x =

        16.00

y =

        27.71
```

Look back at the geometry of the situation shown in Figure 4-16 for which we have written the functions in Figure 4-17 and Figure 4-18. How could you test the results for the projections along x and y? Here is one way:

---

5. The story in this paragraph is not 100 percent true. If you call a function named **myFunction** that is contained in a file named **myFunctionFile**, then you may call it from the Command Window by >> myFunctionFile. This would be bad style in MATLAB because it is so prone to causing user mistakes; in this book we will pretend that such usage is not possible.

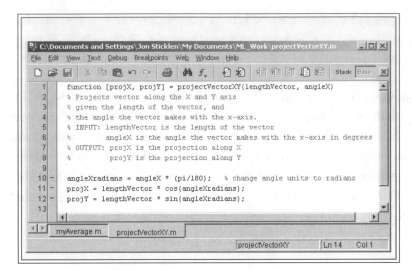

Figure 4-18: Modification of `projectVectorAlongX` to Return Projections Along Both X and Y Axes

```
>> theCheck = (x^2 + y^2)^0.5
theCheck =
          32.00
```

What should the value of **theCheck** be? Solving a problem in MATLAB can be fun and entertaining, but in the end, it is turning a computational crank. You have to find ways when you are using computational tools to check the results those tools give you. The real engineering challenge when you use a tool like MATLAB is to understand the results you obtain and to have confidence that the results are correct.

1. **A MATLAB function is insulated from the outside because it has its own workspace. The variables in a function's workspace are available only to the owning function.**

2. **When a MATLAB function is called, a linkage is made between the actual parameters in the call and the formal parameters in the function definition.**

3. **The number of actual input parameters must be the same as the number of formal input parameters. This is the same for the number of actual output parameters and formal output parameters.**

4. **The first line of a MATLAB function begins with the keyword function, and following that, the rest of the first line looks like an assignment statement.**

Synopsis for Section 4-3

4-4   Testing and Debugging MATLAB Script and Function Files

One certainty about building MATLAB script files or function files is: *You will make mistakes as you solve problems with MATLAB*. No one writes correct code all the time. Learning to use MATLAB to your best advantage includes learning how to identify your coding mistakes and how to correct them. In this section, we will cover the basics of using MATLAB tools for debugging script and function files.

There are two bugs commonly encountered in MATLAB programming (or any other programming environment):

1. Errors result from incorrect applications of MATLAB rules for legal commands. These are called **syntax errors**. Think of these errors as analogous to errors in grammar when you use English to communicate. For example, saying "The solid rocket booster in the Shuttle are vulnerable to low temperatures" is incorrect English grammar usage because the subject of the sentence (booster) is singular while the verb (are) is plural. Even in cases of bad grammar usage, if you heard the sentence you would probably still understand it. But in MAT-LAB (or any programming environment), breaking the rules by which commands are composed will not be understood and will result in a syntax error. A syntax error causes MAT-LAB to abort the computation it was asked to perform.

2. Errors result from incorrect logic. These are called **runtime errors**.[6] The term runtime means during the time MATLAB is executing. Runtime error is partially descriptive: the error takes place while MATLAB is running. What is left unsaid is that a runtime error is not an error of language use. You think you know how to solve the problem, correctly convert your understanding of the problem solution into MATLAB syntax, and MATLAB dutifully does what you have told it to do. The flaw is not in your use of MATLAB; the flaw is in your understanding of how to solve the problem.

To illustrate a runtime error, consider the following example. Suppose you asked me for directions for making a hot fudge sundae and I told you, "First, place a gallon of ice cream in the oven at 350° and bake for two hours. Remove from the oven and pour hot fudge on top." Nothing is grammatically wrong with my instructions. But if you

---

6. This error is also called a **logic error** or a **semantic error**. We will use the term **runtime error**.

follow them, you will have hot fudge soup and not a hot fudge sundae. *Computer programs are directions for solving a problem.* You can write a function or script file using legal MATLAB syntax and still have an incorrect answer for your problem because you do not understand how to solve it.

4-4.1   Syntax Errors

Syntax errors are easy to find: you run the function or script and if MATLAB cannot understand what you are asking it to do, then MATLAB will not complete your computation, i.e., it will go into a *break*. For example, suppose you are adding two numbers in the Command Window, but you forget to include the addition operator:

```
>> x = 32; y = 13;
>> z = x y
??? z = x y
            |
Error: Missing operator, comma, or semicolon.
```

MATLAB is telling you that it cannot understand the assignment command you have given it. You have not followed the rules of syntax for forming MATLAB commands. You cannot have two variables in an expression and not tell MATLAB what to do with the two variables. MATLAB's "informative message" in this case is understandable. But remember, MATLAB reacts with a break when it does not understand what you want to happen. Sometimes, the message that MATLAB gives you will seem obscure because the real syntax error is at the head of a chain of commands, and MATLAB is reacting to the tail of the chain. Sometimes it will become a detective job to figure out where the syntax error is located.

Let us take the same operation (adding two numbers) and consider what happens when you make the same syntax error (forgetting the addition operator). However, instead of this being a command executed in the Command Window, the command is in a function. Figure 4-19 shows our target function. Implement **sumTwoNumbers** and execute the following command:

```
>> aSum = sumTwoNumbers(3,2)
```

Here is MATLAB's response:

```
??? Error: File: C:\ML_Work\sumTwoNumbers.m Line: 6 Column: 15
Missing operator, comma, or semicolon.
```

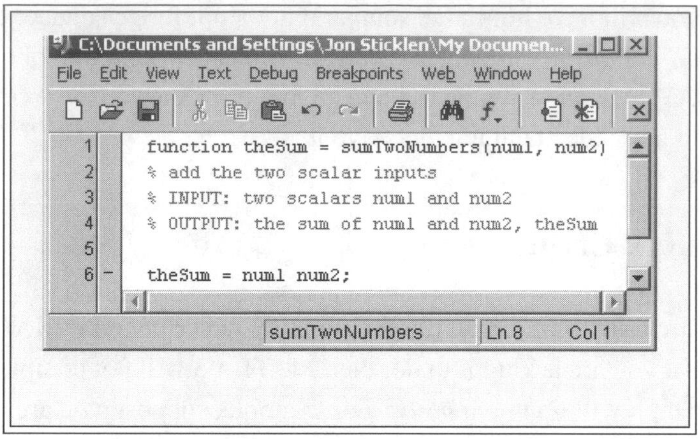

Figure 4-19: Simple Function with Syntax Error

Notice the difference in the messages that MATLAB gives this time. If you execute a command that has a syntax error directly in the Command Window, then MATLAB tells you what it thinks is the problem and tries to point you where it thinks the problem lies. If a command with a syntax error is executed as part of the execution of a function (or a script file), then MATLAB gives the same content message for what it thinks the problem is, but now MATLAB tells you in what file the problem took place, on what line the problem was, and even the column of that line where the problem is. The underline of the file name, line number, and column number is not for emphasis. It is a hot link. Click on the hotlink, and your editor window will open to the file, line, and column of the syntax error.

There is good news and bad news. The good news is syntax errors are usually correctable, provided you know MATLAB programming rules and constructs. Like making a typo in a term paper, once someone proofreading your term paper points out a typo to you, then it is easy to fix provided you know English. In writing MATLAB script files and functions, MATLAB plays the part of the proofreader for you when you run the script or function. If English is not your first language and you have not yet learned it well, then fixing a typo can be a problem as would be for MATLAB. You have to know MATLAB syntax to understand and correct syntax errors.

4-4.2   Runtime Errors

That is the good news: Syntax errors are relatively easy to find and correct. The bad news is runtime errors exist. They are harder to correct than syntax errors because they are harder to find. Why? Because the root problem, if there is a syntax error in a function, lies in your use of MATLAB, which MATLAB itself will recognize and locate for you. But if the problem lies in your understanding of the directions for a solution you are working on, then MATLAB is in the dark. You must find and correct your runtime error. Happily, MATLAB gives you debugging tools to help.

We will start with debugging a single function file. Later, we will add more debugging capability to help if we have multiple function files we hope to debug. Assume that you have located and fixed any syntax errors in the single function file. The steps we have to step through are as follows:

1. Check the function to see if it is working the way we intend.
2. If not, then identify where the problem is in the function.
3. Fix the problem.

This is just common sense put into a formal list. If you constructed a swing set for your younger sister, you would not be done when the last bolt was tightened. You would test the swing set, and if one of the legs of the swing set fell off, you would know that you had more work to do. You would find out what was wrong and correct it.

We can push the swing set analogy and ask the question, "How would you test the swing set?" We need to elaborate what we mean in Step 1 above. The starting point is to understand what you expect the swing set to do. You probably do not expect it to support two orangutans, but you do expect it have a margin of safety for holding kids vigorously swinging. Converting to language a little closer to computer-eze, we need to know what the swing set is capable of to test it. We could limit the input to the swing to be the weight of an average adult and allow for a margin of safety for weight of an average child using the swing set.

In addition to the maximum weight of the swinger, the maximum height the swinger will swing should be considered, i.e., the higher the swing goes, the more stress on the swing set. At this point, we have decided that to test the swing set, you will use maximal values for two important inputs for the swing: the person's weight using the swing and the height that the person will go in the swing.

Having thought this through, you can test the swing set. Using yourself as an average adult, you sit in the swing and test it, swinging to some maximum height that you think your younger sister is likely to go. If the swing set does not fall apart with your test, you are home free, and ready to release the swing to your sister. We followed this testing procedure through step by step, but if you consider the path we followed, you will see the steps are intuitively what you would have done.

Transfer the procedure for testing the swing set to testing a MATLAB function. To restate, a function is a computer function characterized by its inputs, outputs, and how it performs the computational mapping from input to output. Applied to a computer function, *what it's supposed to be capable of* translates to *given a specified input, does the function compute and return the correct output*? We do not need to consider the third characteristic (the computation going on inside the function) unless examination of the function output given a specific input shows that the function has a runtime error. And that is the link back to the second type of error, the one we are trying to nail down now.

There are a number of somewhat complex issues in the simple phrase *given a specified input, does the function produce and return the correct output?* If our goal is to be sure a function is correct, then which input should we test it with? The total set of all possible inputs might be infinitely large. We cannot test them all. If we cannot test all possible inputs, is there another approach? Yes. Researchers in software engineering have developed methods to prove that a given function does what it is supposed to. Such proofs of correctness for software programs have been developed for small and mid-sized programs so far although research continues. In this book, we will not deal with correctness proofs.

That still leaves us with the question, "How do we test a function?" More specifically, what inputs should we test a function with? Suppose we go back to the case with the swing set. In the swing set's physical system, we used the maximal values for the two inputs and concluded if the swing did not fall apart on the maximal inputs, then it met its function. Think about this for a minute. Suppose a fifty-pound, eight-year-older were to sit in the swing at rest. What is it that makes us think that the swing will not fall apart under that input? The answer is that for common mechanical systems, if they work for the maximum loading conditions, then they will work for smaller loads, too. That is, under less than maximal loads, they will stay together.

The problem for software systems, in particular software functions, is that no input type is directly analogous to maximal loads for mechanical systems. We have to search at a higher level of abstraction to find the analogy. The maximal load for a physical system is a special case out of all the loadings we might put on the system. That's the clue. What we want to include as test inputs to a function are values that stand out somehow, ones that test the system's limits.

For example, suppose we need to test a function with one argument, an angle, and that it was built largely from trig functions. Suppose further that the only trig functions we used were sine and cosine. The sine and the cosine functions have ranges that span

$[-1, 1]$. Values for the angles that produce a -1 or a 1 for the sine or the cosine are $\left(0, \dfrac{\pi}{2}, \pi, \dfrac{3\pi}{2}\right)$. If we choose this set of angles as our test inputs, we will be exercising what you might call the far points for our function. Determining the input values to test a function can be difficult. We will return to the topic of test input sets after we have developed more programming machinery. For now, you should use the idea of far points when you test your functions.

Before we move to a concrete example, let us consider one last point. Testing can never prove a function is correct – that it will always work – unless all possible inputs are included in the test set. The purpose of function testing by supplying some selected inputs is to increase confidence the function will work the way we intend it to work. Keep that in mind.

Let us renew the problem of writing a MATLAB function that takes inputs of magnitude $|\vec{r}|$ and angle ϕ with the x-axis for a vector $\vec{r}$, and which computes two outputs: the projection of $\vec{r}$ along the x-axis and the projection of $\vec{r}$ along the y-axis. A sketch of the problem was shown in Figure 4-16. We solved this problem; a correct solution is shown in Figure 4-18. Suppose, though, that we incorrectly remembered the numerical factor to convert degrees to radians and have produced the function file shown in Figure 4-20. In line ten of this function file, the conversion factor for degrees to radians should be 180 and not 360 shown in Figure 4-20. We must use MATLAB's debugging tools to help locate the error.

Bring up the MATLAB editor, and open the file you created earlier named **projectVectorXY.m**. Two lines have changed: line one with the function name, and line ten with the conversion factor for degrees to radians. Use **SaveAs** to create a new file named **FLAWED_projectVectorXY.m**.

The first step in testing a function is to determine if the function is working correctly on a selected set of inputs. Above, we argued that a reasonable set of test inputs for the ϕ input, given that the function uses sine and cosine internally, is $\left(0, \dfrac{\pi}{2}, \pi, \dfrac{3\pi}{2}\right)$. That still leaves the $|\vec{r}|$ input. We developed the informal idea of far points when we talked earlier about the test input set for ϕ. The same idea, applied to the length of $\vec{r}$, could lead us to using the two values $(0, 1)$ for the test input set for $|\vec{r}|$. Why? The 0 is a good candidate for a test value for $|\vec{r}|$ because the scalar value 0 can cause unexpected errors in mathematical functions. The other far point for $|\vec{r}|$ that we can use is the

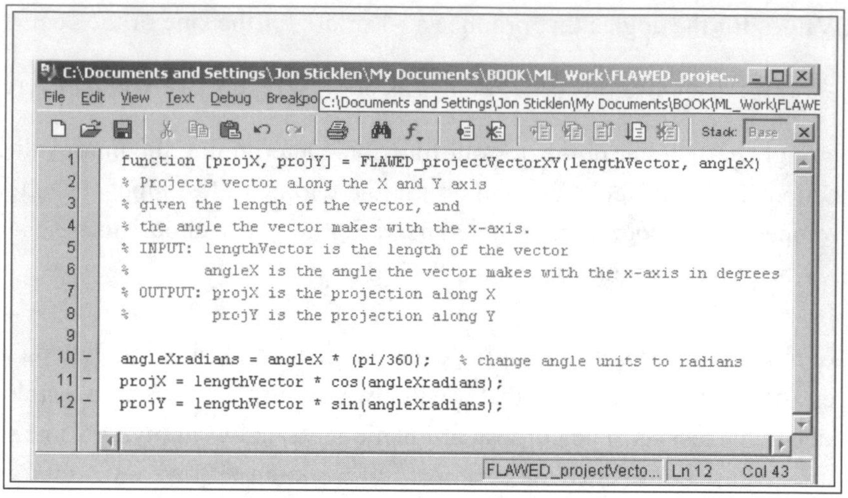
C:\Documents and Settings\Jon Sticklen\My Documents\BOOK\ML_Work\FLAWED_projec...
File Edit View Text Debug Breakpo C:\Documents and Settings\Jon Sticklen\My Documents\BOOK\ML_Work\FLAWE
Stack: Base
function [projX, projY] = FLAWED_projectVectorXY(lengthVector, angleX)
% Projects vector along the X and Y axis
% given the length of the vector, and
% the angle the vector makes with the x-axis.
% INPUT: lengthVector is the length of the vector
%   angleX is the angle the vector makes with the x-axis in degrees
% OUTPUT: projX is the projection along X
%   projY is the projection along Y
angleXradians = angleX * (pi/360);   % change angle units to radians
projX = lengthVector * cos(angleXradians);
projY = lengthVector * sin(angleXradians);
FLAWED_projectVecto... Ln 12 Col 43

Figure 4-20: Flawed Version of `projectVectorXY`

value 1. The value 1 is one of an infinite set of values for $|\vec{r}|$ that are typical. The idea with the test set of $|\vec{r}|$ is that we are choosing one special point (the value 0) and one typical point (the value 1). If the function will work right on these two values, then we have some confidence it will work right on all values of $|\vec{r}|$. We have then four test values for ϕ and two test values for $|\vec{r}|$. To test our function on the inputs we have selected for $|\vec{r}|$ and ϕ, we will have a total of eight (4×2) input pairs.

Make a table similar to the table in Table 4-2, which lists in its first two columns the input values for $|\vec{r}|$ and ϕ, in its second two columns the expected and actual values for the *x* projection of $\vec{r}$ given the row inputs, and in its second two columns the expected and actual values for the *y* projection given the row inputs. This sounds complicated but is not because we are systematically writing down the set of inputs we want to use to test the function and logging what we expect the function to produce and what it actually produces. The expectation and the actual result should match if the function is working right.

We want to find runtime errors, errors that are due to the MATLAB function having incorrect instructions in it. If we follow the instructions in **`FLAWED_projectVectorXY`** to calculate the expected *x* and *y* projections for Table 4-2, then we will not be testing for incorrect instructions in the code. To get the expected values to use in looking for runtime errors, you must use means that are separate from the code you have written. For testing the function **`FLAWED_projectVectorXY`** we can use the diagram of Figure 4-16, and augment it to show the *y* projection as well as the *x* projection.

Table 4-2: Test Table for Function
FLAWED_projectVectorXY.m

$|\vec{r}|$ (m)	ϕ (rad)	expected x (m)	actual x (m)	expected y (m)	actual y (m)
0	0				
0	$\dfrac{\pi}{2}$				
0	π				
0	$\dfrac{3\pi}{2}$				
1	0				
1	$\dfrac{\pi}{2}$				
1	π				
1	$\dfrac{3\pi}{2}$				

The augmented diagram showing projections along both axes is shown in Figure 4-21. Using only this diagram, without a calculator, you should be able to fill in the expected values in Table 4-2. First, consider all the rows in which $|\vec{r}| = 0$. If the length of vector $\vec{r}$ is 0, then its projection on the x and y axes is going to be 0; so, we know the expectations for half of the rows in Table 4-2. That leaves the rows in which $|\vec{r}| = 1$. If $\vec{r}$ is along the x-axis, then its projection along x is its length and its projection along y is 0. If $\vec{r}$ is along the negative x-axis (that is, if $\phi = \pi$), then its projection along x is the negative of its length, and its projection along y is 0. You can use this same reasoning process and the diagram in Figure 4-21 to fill in the remaining expectation values in Table 4-2.

The two columns in Table 4-2 for actual values are for recording the real output values from **FLAWED_projectVectorXY.** To get them, you run **FLAWED_projectVectorXY** in MATLAB with the specified input.

Complete filling in all values in Table 4-2. You will find that the $|\vec{r}| = 0$ inputs (all four of them) produce expected output values for the x and y projections. You will find three of the four values for ϕ produce unexpected outputs when $|\vec{r}| = 1$. In general, you know two things:

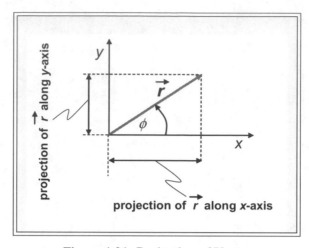

**Figure 4-21: Projection of Vector on
x- and y-axes**

1. You know that **FLAWED_projectVectorXY** does not
    produce expected output for all of the test input values.
    Hence, you know it has a runtime error.

2. You know which of the test input pairs produce the unex-
    pected output. This is important. You need to know the input
    pairs that produce incorrect output because we will use those
    inputs in probing to locate and fix the runtime error.

We will take one of the input pairs that produces incorrect results, and use it as a test
probe for $|\vec{r}| = 1$, $\phi = 180°$. From your work to fill in the missing values in Table 4-2,
you know that for the test probe, the expected and actual outputs produced are shown
in Equation 4-3:

$$\vec{r}_{anticipatedAlongX} = -1$$
$$\vec{r}_{actualAlongX} = 0$$
$$\vec{r}_{anticipatedAlongY} = 0$$
$$\vec{r}_{actualAlongY} = 1$$

(EQ 4-3)

Earlier, in the discussion of Figure 4-14 on page 4-100, you learned to use the
MATLAB debugging facility to pause a function while it was running. We are going to
use that machinery to find the runtime error in **FLAWED_projectVectorXY**.

Bring up **FLAWED_projectVectorXY** in your MATLAB editor window. Under
the **Debug** menu, select *Stop if Errors/Warnings ....* Then on the *Errors* tab, click
Always stop if error. *Stop If Error* only refers to syntax errors: MATLAB on its own
cannot locate a runtime error for you. But it is always good practice when you are

testing/debugging a function to turn this facility on. The effect is that MATLAB will pause and wait for you to do something when it locates a syntax error, and that means that the function's workspace in which the break took place will be available to you.

Look back at the code for **FLAWED_projectVectorXY** in Figure 4-20. The problem-solving context is that we know that **FLAWED_projectVectorXY** produces unanticipated (wrong) output given the input of $|\vec{r}| = 1, \phi = 180°$. MATLAB has detected no syntax error. It is our job to find and correct the runtime error.

Execute the following in your Command Window:

```
>> [x,y] = FLAWED_projectVectorXY(1, 180)
```

This is the result that MATLAB will hand you is:

```
    x =

            0.00


    Y =

            1.00
```

The correct answer is $x = 1, y = 0$.

Set a break point at line ten of **FLAWED_projectVectorXY**. Do this by clicking over the dash mark to the right of the line number. Your editor window will look like Figure 4-22.

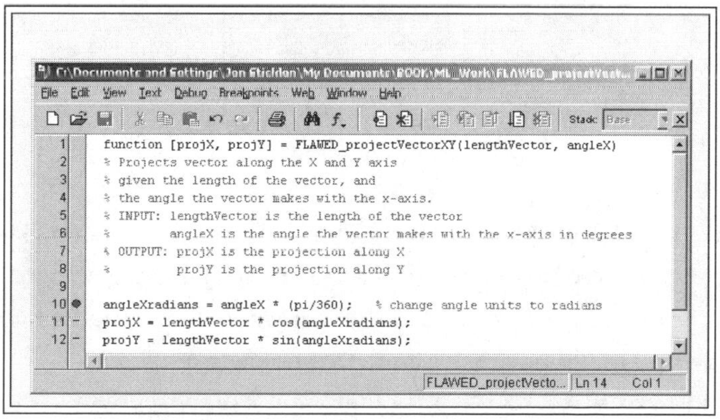

**Figure 4-22: FLAWED_projectVectorXY
with Break Point Set**

Why at line ten? Because we know something is haywire in **FLAWED_projectVectorXY**, the function only has three lines of working code, and line ten is where the working code starts. By pausing the function before the point at which the working code starts, we will be able to take control of function execution, and see what values are being computed internally in **FLAWED_projectVectorXY**.

Execute the command again to run **FLAWED_projectVectorXY** with arguments $|\vec{r}| = 1$, $\phi = 180°$. Because you have a break point set at line ten, the function pauses before executing line ten. Now, you can examine the workspace of **FLAWED_projectVectorXY**, as shown in Figure 4-23. Double-click **lengthVector** in the workspace to see its current value, which should be the value you handed **FLAWED_projectVectorXY** as the first input: the number 1. Verify the second argument you input to **FLAWED_projectVectorXY**: The input variable **angleX** should have the value 180. To make the process explicit, our game is one of matching what we expect values to be inside **FLAWED_projectVectorXY**, and matching those expectations against what those values really are. When our expectation does not match the case inside **FLAWED_projectVectorXY**, then we may have identified the source of our runtime error.

**Figure 4-23: Workspace
Before Line Ten**

Next, pull down the **Debug Menu** in the editor window. Select Step. The action that MATLAB takes is to execute one line of code (line ten) and pause before executing the next line (line eleven). After line ten is executed, then a variable named **angleXradians** should be created in the workspace of **FLAWED_projectVectorXY**: Verify that your Workspace window looks like that in Figure 4-24. Look at **angleXradians** in the Workspace window to get its actual value, or hold your cursor over **angleXradians** in the editor window.

**Figure 4-24: Workspace
Before Line 11**

Note that **angleX** has the value $180°$. The value of **angleXradians** should be π radians; 3.1416 is our expected value for **angleXradians**. But the actual value for **angleXradians** is 1.5708. This result tells us we have a run-time error in line ten.

Having pinpointed the line of the runtime error in **FLAWED_projectVectorXY**, we need to correct the error. The comment for line ten indicates this code line is supposed to convert the angle to radians. When line ten was written a mistake was made in the conversion factor, and that problem can be corrected by changing the conversion factor from 360 to 180.

Make that change in the command for line ten. The red stop sign in the edit window has become grayed, which indicates a code change has been made. To proceed, you need to re-save the function definition of **FLAWED_projectVectorXY**. From the **File** menu, select **Save**. MATLAB will warn you that saving will abort the break. Click OK. You are back at the regular prompt in the Command Window, and ready to see if correcting the conversion factor in line ten fixed **FLAWED_projectVectorXY**, or if there are other runtime errors that must be found.

Running through all the test inputs with the correction will show you that only line ten had a problem. Having fixed that problem, you have completed testing and debugging for **FLAWED_projectVectorXY.**

The example of testing and debugging for **FLAWED_projectVectorXY** was tedious even though you knew what was wrong with **FLAWED_projectVectorXY** before we started the example. The point here is not **FLAWED_projectVectorXY** but rather the general process you follow in testing and debugging a function. The key to effective testing and debugging is to be systematic. That especially applies to selecting the test inputs you will use to test the function, and figuring out the expectations you have for the expected outputs for the function applied to the test set before you use MATLAB to output the actual values. If you find a mismatch between your expectations for the output variable and what your function actually computes, you have a runtime problem. You must find the problem inside the function, and the easiest way to do that is to methodically figure out which line(s) of code is home for the runtime error you have found. Once having found the runtime error, you job becomes one of fixing the problem(s).

Earlier, we talked about the importance of making MATLAB variable names be descriptive and putting comments in code to help understanding. Given the problem **FLAWED_projectVectorXY**, the name of the variable **angleXradians** probably would have been enough to remind you what the variable should hold: The value of the angle the vectors makes with the *x*-axis in radians. The value of the angle in degrees was input to the problem in the input variable **angleX**. Without meaningfully named variables and informative comments, you would probably have a more difficult time developing your expectations for what each line of code should produce and, hence, a much harder road to follow to find and correct runtime errors.

Later in this book, you will learn more capabilities of the MATLAB debugging tools. But with what you have learned in this chapter, you will be able to find most runtime errors.

1. **To test a function in MATLAB, you must identify the set of inputs you will use for the test.**
2. **You should determine, without using MATLAB, what you expect each of the test input sets to produce in terms of MATLAB outputs.**
3. **Comparing what you expect for the test inputs to what MATLAB produces enables you to identify runtime errors in the function you write.**
4. **If you find a runtime error in a function, then you must identify and correct the line(s) of code in the function that are causing the runtime error.**

Synopsis for Section 4-4

4-5   Problem Sets for Saving Your Work

When you write any function or script, include appropriate comments.

4-5.1   Set A: Nuts and Bolts Problems for Saving Your Work

Problem 4-A.1 (Section 4-1)

By default, in which folder does MATLAB save files? Which folder will MATLAB search first to load or run saved files? Write your answer in a concise paragraph or two.

Problem 4-A.2 (Section 4-1)

What is the purpose of the Current Directory Window? What does Current Directory mean in MATLAB? Write your answer in a concise paragraph or two.

Problem 4-A.3 (Section 4-1)

In Section 4-1, you created a new folder named **ML_work** under **My Documents**. Create a new folder under **ML_work** and name it **Ch4_Saving**. Make **Ch4_Saving** the current directory.

Put your solutions for the following problems in **Ch4_Saving**.

Problem 4-A.4 (Section 4-2)

The area and the perimeter of a circle are found with the following expressions:

$$circleArea = \pi r^2$$
$$circlePerimeter = 2\pi r$$

In the MATLAB Command Window, set **r** to be 4 meters. Record the units as a comment in the command.

Now write a MATLAB Script file called **circlePerimeterArea** that computes the area and the perimeter for a circle of radius **r**. Put appropriate comments into your script file indicating the following:

- Its file location

- Its purpose

- Any variables which are assumed to be set when the script file runs

- The units of the assumed variables

Use Figure 4-4 as an example. All lines but the last end in semicolons. The last line produces an echo of the variables the script file produces because it has no ending semicolon. Your script file should be similar to the one shown in Figure 4-4.

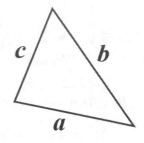

Save your work.

Run **circlePerimeterArea**.

Change the radius to 7 meters in Command Window and rerun the script file. Test your results.

Problem 4-A.5 (Section 4-2)

Suppose you know the lengths and the angle between two adjacent sides of a triangle, as shown in the sketch to the right. You can find the area of the triangle area from the following relation:

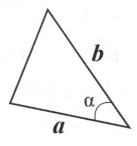

$$triangleArea = \frac{1}{2}ab\sin(\alpha)$$

In the Command Window, set side length **a** to be 5 ft, side length **b** to be 6 ft, and angle α to be 23°.

Enter **edit** to open MATLAB editor.

Write a script file to compute the area of the triangle using the relation above and store it in the variable **SAS_TriangleArea**. (The name comes from Side Angle Side computed area.)

The values of **a, b**, and α should not be set in your script file. Rather the script file will use the values of the variables you set in the Command Window. The trig built-in functions in MATLAB expect angles to be in radian units.

Store your script file giving it the name **computeSASarea**.

Run your script file, and test the results you obtain.

Reset **a, b**, and α to be 12 feet, 2 feet, and 78°, respectively.

Run your script file again, and test your new results.

Problem 4-A.6 (Section 4-2)

Suppose you know all the side lengths of a triangle, as shown in the sketch to the right. You can find the triangle area using Heron's formula:

$$triangleArea = \sqrt{S(S-a)(S-b)(S-c)}$$

S is the semi-perimeter and has the following value:

$$S = \frac{a+b+c}{2}$$

In the Command Window, set variables **a**, **b**, and **c**, to be 6 feet, 7 feet, and 3 feet, respectively.

Build a script file that computes the area of the triangle defined above, and stores the result in a variable named **SSS_TriangleArea**. Include a comment that explains your script file's purpose.

Save your script file; name it **computeSSSarea**.

Run your script file and test your results.

Reset **a**, **b**, and **c** to be 12 feet, 10 feet, and 15 feet, respectively. Run your script file again, and test your results.

Problem 4-A.7 (Section 4-2)

Using the MATLAB script files
computeSASarea.m and
computeSSSarea.m that you wrote in Problem
4-A.5 and Problem 4-A.6, find the area of the
shape shown in the sketch to the right.

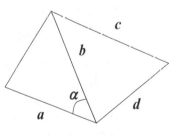

First, in the Command Window, set the variables **a**,
b, **c**, **d**, and α to be 5 feet, 6 feet, 7 feet, 11 feet,
and 23°, respectively.

Create a new script file called **twoTriangles.m**
whose purpose is to call **triAreaSAS** and **triAreaSSS** and add the results to determine the total area of the object in the sketch above. Set a variable named **sumBothTriangles** to hold the resulting total area.

Be careful with the names of variables you used in **triAreaSAS** and **triAreaSSS**. You may have to make changes in **triAreaSAS** and **triAreaSSS** to ensure consistent variable naming.

Put comments in **twoTriangles** that explains the purpose of your script file, and the names and meanings of variables used.

Run **twoTriangles** to test your result.

Now reset **a**, **b**, **c**, **d**, and α to be 3 feet, 5 feet, 7 feet, 15 feet, and 32°, respectively. Run **twoTriangles** again and test your new result.

Problem 4-A.8 (Section 4-2)

Consider the parallelogram in the sketch to the right.

In the Command Window set variables **a**, **b**, **c**, **d**, **e**, α, and β to be 7 feet, 8 feet, 5 feet, 6 feet, 3 feet, 66°, and 85°, respectively.

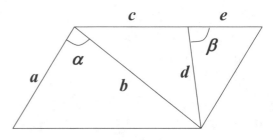

Write a new script that calculates the area of the parallelogram. Use the scripts you wrote for Problem 4-A.5 and Problem 4-A.6. Name your calling script file **areaParallelogram**.

Put the result for the total area of the parallelogram in a variable named **sumAreaParallelogram**.

Keep the variables' names consistent across all scripts in your solution. You may need to make changes in scripts you wrote earlier.

Run **areaParallelogram** and test your result for the current settings of the problem variables.

Reset **a**, **b**, **c**, **d**, **e**, α, and β to be 3.5 feet, 4 feet, 2.5 feet, 3 feet, 1.5 feet, 66°, and 85°, respectively.

Run **areaParallelogram** again, and test your new result.

You have completed a number of problems centered on writing MATLAB scripts. One of the points you need to learn is the difference between MATLAB user-defined functions and MATLAB user-written scripts. Ensure that in each case of writing a script, you defined variables used in scripts in the global workspace and any variables generated in scripts in the global workspace. This in turn leads to a need to be certain that all scripts used in a given problem employed consistent variable names. That, in turn, can lead to a need to modify existing scripts with each new problem.

In the remainder of this book, when you are asked to solve a problem, you should write a user-defined function unless you are explicitly told to write a script.

Problem 4-A.9 (Section 4-3)

Answer the following questions in the context of workspaces:

> - What do the terms *global workspace* and *local workspace* mean?
> - How can you retrieve the value of a variable used in a function from the Command Window?
> - Can you edit a function workspace from the Command Window in the normal mode of operation?
> - Can you edit a script workspace from the Command Window in the normal mode of operation?

Answer all of the questions in one or two concise paragraphs.

Problem 4-A.10 (Section 4-3)

Suppose you write the following functions (in 2 separate DOT-M files) for finding the sum-of-squares of two numbers:

```
ss1:   function [sqSum] = squareSum(a,b)
ss2:      sqA = square(a);
ss3:      sqB = square(b);
ss4:      sqSum = sqA + sqB;
```

```
sq1: function [sq] = square(a);
sq2:    temp = a;
sq3:    sq = temp^2;
```

Then, you test **squareSum** by running it from the MATLAB Command Window:

```
cw1:   >> clear;
cw2:   >> temp = 3^2 + 5^2;
cw3:   >> finalValueSqSum = squareSum(3,5)
```

Fill in the following table with the values of variables while MATLAB is running. Remember each function has its own workspace. Enter *Unk* if a variable is undefined (not in the workspace currently active).

Problem 4-A.11 (Section 4-3)

Write a MATLAB function that takes an angle in degrees and converts it to radians. Use this conversion relation:

location of probe	a	b	sqA	sqB	sq	temp	finalValueSqSum
in Cmnd Wndw: **after cw2, before cw3**							
in **squareSum**: after ss1, before ss2							
in **square**: 1st call, after sq3							
in **squareSum**: after ss2, before ss3							
in **square**: 2nd call, after sq3							
in **squareSum**: after ss3, before ss4							
in **squareSum**: after ss4							
in Command Window: **after cw3**							

$$angleInRadians = angleInDegrees \times \frac{\pi}{180}$$

Name your function **deg2rad**. (The name comes from "convert degrees to radians.") A function must minimally include comments indicating the function's purpose, input variables used by the function, and output variables used by the function.

Test your function with the following equation:

```
>> angleRad = deg2rad(45)
```

Problem 4-A.12 (Section 4-3)

Write a MATLAB function that takes an angle in degrees and returns the cosine of the angle. In your function, do the following:

- Call **deg2rad** of Problem 4-A.11 to convert the angle to radians.
- Find and return the cosine of the angle.

Name your function **cosDeg.m**.

Call your function from the MATLAB Command Window to test its operation.

```
>> cosAngle = cosDeg(74)
```

Include appropriate comments in your function.

Problem 4-A.13 (Section 4-3)

Modify your solution to Problem 4-A.12 so your function returns the angle in radians and its cosine.

Use *SaveAs* under the file menu in the MATLAB editor to create a copy of **cosDeg**; name the new copy **radCosD**. Test your function by calling it from MATLAB Command Window:

```
>> [angleRad, angleCos] = radCosD(74)
```

Problem 4-A.14 (Section 4-3)

Refer to the sketch to the right. For right triangles, the tangent of angle α is defined as the ratio of the opposite side length to the adjacent side length. The equation for the sketched triangle is:

$$Tangent(\alpha) = \frac{a}{b}$$

Write a MATLAB function that takes sides **a** and **b** (both in meters) and returns the tangent of α, as shown in the sketch. The order of the inputs to your function is important.

Name your function **tanTriangle.m**.

Test your function from the MATLAB Command Window:

```
>> tanAlpha = tanTriangle(5,3)
```

Problem 4-A.15 (Section 4-3)

Modify your solution for Problem 4-A.14 so the function returns the tangent and the value of the angle (in degrees).

Name your function for this problem **tanAndAngle**. Test your function by calling it from the Command Window and testing the result, use the following:

```
>> [tanAlpha, alphaInDeg] = tanAndAngle(5,3)
```

Problem 4-A.16 (Section 4-3)

Write a MATLAB function that takes five scalar inputs and returns the arithmetic and geometric mean of the five numbers.

The mathematical definitions of the simple arithmetic mean of N numbers (a_1, a_2, ..., a_i, ..., a_N) and the geometric mean of the same N numbers are these two equations:

$$arithmeticMean = \left(\sum_{i=1}^{N} a_i \right) \times \frac{1}{N}$$

$$geometricMean = \left(\prod_{i=1}^{N} a_i \right)^{\frac{1}{N}}$$

(Look up geometric mean using a search engine such as Google if you don't understand the $\prod$ symbol.)

Name your function **arithGeomAvg**. Test your function by executing the following:

```
>> [arithM, geomM] = arithGeomAvg(5,3,8,5,7)
```

Problem 4-A.17 (Section 4-4)

Which type of error is easier to detect a syntax or a runtime error? Why?

Write your answer in one or two concise paragraphs.

Problem 4-A.18 (Section 4-4)

Identify and correct the syntax errors in the following commands. After you believe you see the problem, and the correction you need, then type the original error and your correction into the Command Window to see if you are correct.

Part A   `y = 5(8 - 5)^2/7`

Part B   `tan45 = sin[45(pi/180)]/cos[45(pi/180)];`

Part C   `sinSquare45 = sin^2(45*pi/180)`

Part D   `EricAge + 7 = 41`

Part E   `(4.7) =< (10/4)`

Part F   `(5/2) = (10/4)`

Problem 4-A.19 (Section 4-4)

The total interest on an initial investment of M dollars compounded annually at r% interest is given by

$$interestEarned = M\left[\left(1 + \frac{r}{100}\right)^k - 1\right]$$

Use the following function to calculate the total interest and overall balance:

```
function [interest, balance] = bankAccount(M, r, k)
% This function calculates the account balance
% compounded annually over the interest rate, after
% a specified number of years
% Inputs: Initial investment: M (dollars)
%         percent interest rate: r (percent)
%         number of years: k (yrs)
% Outputs: interest earned: interest (dollars)
%          bank account balance: balance (dollars)
   interest = M((1+r)^k-1);
   balance = M+interest*k;
```

Create a copy of the function above using the MATLAB editor and save it in your connected directory.

Find and fix the syntax error(s).

Detect the runtime error(s) and fix them.

Problem 4-A.20 (Section 4-4)

The area of a triangle for which all side lengths are known can be found using Heron's formula ...

$$triangleArea = \sqrt{S(S-a)(S-b)(S-c)}$$

S is called the semi-perimeter and has the following value:

$$S = \frac{a+b+c}{2}$$

A novice programmer writes the following function to implement Heron's formula:

```
function area = SSSTriArea(a,b,c)
% This function calculates the area of a triangle
% for which the side lengths are known.
% Inputs: Side lengths of triangle: a (ft), b(ft),
% c(ft)
% Output: Area of triangle: area (sq ft)
  A = sqrt(S)*(S-a)*(S-b)*(S-a);
  S = a+b+c;
```

Create this function and save it in your connected directory.

Find and fix the syntax error(s).

Using the MATLAB debugging tool, find the runtime error(s), and fix them.

Try your function on a right triangle with side lengths 3, 4, and 5 feet using the following command line:

```
>> triArea = SSSTriArea(3,4,5)
```

Calculate this triangle's area by hand as well to test your answer.

4-5.2   Set B: Problem-Solving Using MATLAB User-Defined Functions

Although there will not be reminders in the problems statements below, when you write a user-defined function, you must include appropriate comments.

Problem 4-B.1

You are designing a storage silo for corn. Your design uses a right prism with equilateral triangular bases. In your first try, each side of the triangular end plates has a length of 30 meters and the height of the silo is 10 meters. This silo design holds enough corn to meet requirements.

Make a rough sketch of the first silo design. (If you are unsure what a right prism is, look it up using Google.)

On further consideration, you see that the barn that houses the silo is short on floor space. Suppose that to make the base of the silo have a smaller footprint, you redesign a taller and skinnier silo but with the constraint that it must hold the same amount of corn as the first design. You decide that the redesigned silo should have a footprint only 75 percent of the first design but hold an equal amount of corn.

Make a rough sketch of the second silo design.

Your problem: Determine the dimensions of the second (and final) version of the silo. (Your solution will include the length of a side of triangular end caps and the height of the prism.)

To solve this problem, write the following **script files**:

- **eqTriangleArea**, whose purpose is to calculate the area of an equilateral triangle given the length of a side (in meters)

- **eqTriangleSide**, whose purpose is to compute the side of an equalaterial triangle given its area (in square meters)

- **siloProblem**, whose purpose is to compute the solution to this problem using the two other script files (This is the top-level script file.)

The comments in a script file must include the variables' names that must be set before the script is run and the names of variables set by action of the script. If you don't know the mathematical formula you need to solve this problem, look up equilateral triangle using Google.

Be sure to test your implementation.

Problem 4-B.2

Redo Problem 4-B.1 to the same problem specification as before, but implement your solution with user-defined functions rather than scripts.

Create the following:

      – **eqTriangleAreaFunc** that inputs side length of an equilateral triangle, computes the area of the triangle, and outputs the area

      – **eqTriangleSideFunc** that inputs the area of an equilateral triangle, computes the length of a side the triangle, and outputs the side length

      – **siloProblemFunc** that takes no inputs, computes a solution to the problem, and outputs the solution

Test your implementation.

Write a one- or two-paragraph comparison of using scripts in MATLAB versus using user-defined functions.

Problem 4-B.3

The sketch at the right shows two tugboats pulling a barge in a river. Tugboats A and B pull the Barge with 2,000 and 4,000-lb. forces, respectively. The sketch shows Tugboat A's direction of pull, 35 degrees from the horizontal in the specific case shown. Finally, the sketch shows Tugboat

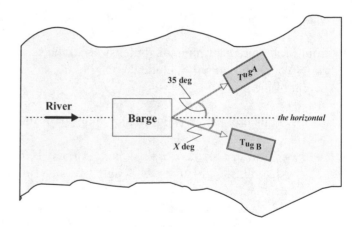

B pulling the barge at an unknown angle, X degrees, in the negative direction from the horizontal.

The goal of the two tugboats is to pull the barge straight down the river, which means along the horizontal. Part of the force exerted by each is along the horizontal while part of the force exerted by each is along an axis that is 90 degrees rotated from the horizontal. In order for the combined pull of the two tugboats to move the barge straight down the river, the pull of Tugboat A in the vertical direction must be as large as the pull of Tugboat B in the vertical but opposite in direction.

Write a MATLAB function that calculates the angle labeled X in the sketch.

Your function should take three inputs: the force exerted by Tugboat A (in lbs), the force exerted by Tugboat B (in lbs), and the angle Tugboat A's pull makes with the horizontal (in degrees). Your function should compute and output the angle that Tugboat B's pull makes with the horizontal (in degrees) to satisfy the goal of moving the boat straight down the river. In addition to the standard comments, include in your function any comments about assumptions you must make to solve this problem. Call your function **angle_TwoTugboats**.

Test your function by executing the following:

```
>> angB = angle_TwoTugboats(2000, 4000, 35)
```

Problem 4-B.4

The transient behavior of RLC circuits (containing resistors, inductors, and capacitors) plays an important role in the design of electronic components. Suppose an open RLC circuit has an initial charge of q_0 (in coulombs) stored within a capacitor. When the circuit is closed the charge flows out of the capacitor, leaving a charge q (in coulombs) after a time t given by the following relation:

$$q = q_0 e^{-\left[\frac{R}{2L}\right]t} \cos\left(\left[\frac{1}{LC} - \left(\frac{R}{2L}\right)^2\right]^{0.5} t\right)$$

R is the resistor in the circuit, L is the inductor in the circuit, C is the capacitor in the circuit, q_0 is the initial charge of the capacitor in the circuit, t is time since the circuit has been closed, and q is the charge remaining on the capacitor at time t. The standard units for R, L, C, q (and q_0), and t are ohms, henries, farads, coulombs, and seconds, respectively.

Write a MATLAB function that takes five inputs for an RLC circuit: R (in ohms), L (in henries), C (in farads), q_0 (in coulombs), and t (in seconds). Your function should compute and output one variable: q, the charge remaining in the capacitor at time t.

Name your function **q_RLC**.

Perform an initial test of your function by executing the following:

```
>> R = 100;       % ohms
>> L = 9.2;       % henries
>> C = 0.0025;    % farads
>> q0 = 1000;     % columbs
>> chargeRemaining_00 = q_RLC(R,L,C,q0,0)
```

Your answer should be the value of **q** at time equal to zero, i.e., 1,000 coulombs. (Why is this a reasonable initial test?)

Now run your function **q_RLC** for **t=0.1**, **t=0.2**, **t=0.3**, and **t=0.4** seconds.

On paper, create a plot of *Charge versus Time* for the five value pairs you have.

How would you characterize the relation between charge remaining and time? Write a concise paragraph describing your characterization.

Problem 4-B.5

A fence is needed around the field sketched at the right. The field consists of a rectangle (**bcef**) and two triangles (**abf** and **cde**). The necessary fence is indicated by the heavy line in the sketch.

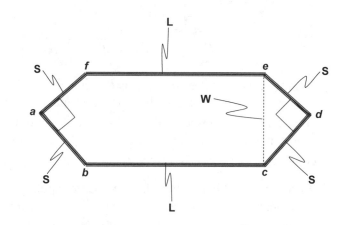

The rectangular part of the field has dimensions **W** (meters) by **L** (meters). The triangles are the same except for a rotation; both are right isosceles triangles with two sides of length **S** and one side of length **W**.

The total area of the field is **A** square meters.

Your job is to calculate the fence cost required for the field. From land records, you know the value of **A**. From measurement, you know the value of **S**. And you are given the cost per linear meter of the fence.

Work out a solution to this problem on paper. Then implement your solution by developing MATLAB functions. Do not do this problem by writing one function. Decompose the problem by writing the functions that compute the following:

1. The length of **W** (meters) given the length of **S** (meters). Name this function **wGivenS**.

2. The area of a right isosceles triangle given the length of one of the isosceles sides (in meters). Name this function **area_rtIsTri**.

3. The side length of a rectangle given its area and width: **length_rec**.

4. Top-level function. The cost of the fence given the area of the field **A**, the length of an isosceles side **S**, and the unit cost of the fence. Name this function **priceFenceTri**.

Once you have your functions written and the syntax bugs deleted, be certain you test your solution with several reasonable inputs.

Problem 4-B.6

Suppose you write the following function to find the square of the ratio of two numbers:

```
function divsq = ratioSquare(x1, x2)

% This function finds the square of the ratio of
% two numbers
% Inputs x1 and x2, simple numbers
% Output divsq is the square of the ratio x1/x2


  divsq = x1/x2^2;
```

Then, you run **ratioSquare** in MATLAB Command Window:

```
>> x1 = 640;
>> x2 = 320;
>> divSquared = ratioSquare(x2,x1)
```

What should the values of **x1**, **x2** and **divSquared** be after the computation is over? What value do you get when you run the function?

Fix any runtime errors in **ratioSquare** and retry.

Problem 4-B.7

You write the following function to add three numbers at a time:

```
function [result] = adding(i1,i2,i3);
% This function returns the sum of its three
% scalar inputs
% Inputs: i1, i2, i3 ...  are three real numbers
% Outputs: result is the sum of the three numbers
  temp = i1 + i2;
  result = temp + i3;
```

Then you call **adding** from the MATLAB Command Window:

```
>> a = 2
>> temp = 3
>> c = adding(a, a, a)
```

When execution of **adding** completes, what is the value of **temp**?

Solve the problem by hand first, then write the function and run it to test your answer.

Problem 4-B.8

You write the following function to find the average of three numbers:

```
function ave = myAverage(x1, x2, x3)
% This function finds the average of three real
% numbers
% Inputs: x1, x2, x3 are all real numbers
% Output: ave is the arithmetic average of the
%     three inputs
  ave = sumInputs/3;
```

You call **myAverage** from the Command Window as follows:

```
>> x1 = 5; x2 = 7; x3 = 2;
>> sumInputs = x1+x2+x3;
>> average = myAverage(x1,x2,x3)
```

What will MATLAB return?

Write a paragraph explaining what is wrong with **myAverage**.

Problem 4-B.9

The solutions for values of x in the quadratic polynomial $y = ax^2 + bx + c$ can be found this way:

$$x_1 = \frac{-b + \sqrt{b^2 - 4ac}}{2a}$$

$$x_2 = \frac{-b - \sqrt{b^2 - 4ac}}{2a}$$

Write a MATLAB function called **qRootsGreaterTen** that inputs coefficients a, b, and c. Your function should return a single output as follows:

- If x_1 and x_2 are real numbers greater than 10, then output **1**.
- Otherwise, output **0**.

(Hint: the square root of a negative number is not a real number. In your function, first compute the values for x_1 and x_2. Then, to compute the output for qRootsGreaterTen, form a logical expression that tests to see if both are greater than 10 and makes certain that the values are real numbers.)

Problem 4-B.10

Automobile transmissions are used to reduce the torque (the turning force on a shaft) while increasing the angular velocity (how rapidly the shaft spins), or vice versa. Consider the simplified transmission model shown in the rough sketch at the right.

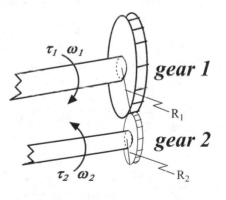

The sketch shows two gears hooked together, or coupled. Real transmissions typically have many more than two gears. Whether simple or complex, systems of gears can be understood by noting the distance that any one gear must rotate. For example, if Gear 1 has a radius half as big as Gear 2, then when Gear 1 completes one complete rotation, Gear 2 will complete only 180 degrees of rotation. This intuition leads to a relation between the angular speed of Gear 1, its radius, and the angular speed of Gear 2 and its radius. Assuming that the teeth on Gear 1 and Gear 2 have the same spacing, the relation is the following:

$$\frac{R_1}{R_2} = \frac{\omega_2}{\omega_1}$$

ω is the angular speed and R is the radius of a gear.

Write a function that inputs the radius of Gear 1, the angular speed of Gear 1, and the desired angular speed of Gear 2. The function should compute and output the radius that Gear 2 to meet the design goal for the gear system (that is, the desired angular speed of Gear 2). Name your function **gear2_rotSpeed**.

Problem 4-B.11

In the sketch to the right, an electric circuit is shown. The circuit consists of a source of electric potential (V volts) that creates an electric current (i amps) through a resistance (R ohms) and some load (like an electric motor). The purpose of the resistance R is to drop the voltage to some desirable level e required to make the load work properly.

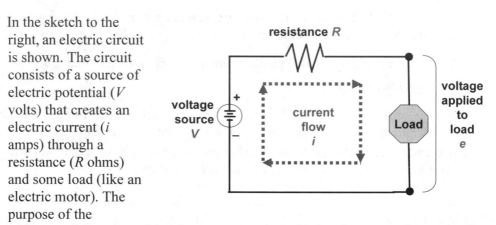

The voltage applied across the load, e, can be determined from this equation:

$$e = V - iR$$

Write a MATLAB function named **resistanceGivenV_I** that computes the value of R given V, i, and e.

Use your function to compute the correct design value for the resistance using these values:

- The electric potential V is 30 volts
- The current flowing i is 5 amps
- The desired voltage e applied to the load is 8 volts

Problem 4-B.12

A container is made up of a cylinder and a cone at the bottom end. The cylinder height and the cone height are equal to the cylinder diameter.

Make a sketch of the container. Be sure to label your sketch and understand the physical situation with which you are working.

Write a MATLAB function that takes a requested container volume as input and returns the cylinder diameter. Name your function **cylDiameter**.

Problem 4-B.13

You are testing the following trig identity:

$$\sin^2(\alpha) + \cos^2(\alpha) = 1, \text{ for all } \alpha$$

You write a function called **pythagorasForUnitCircle** and run it for 0, 30, 60, and 90 degrees. Unfortunately, the program produces large numbers of error messages. You conclude the identity is incorrect. Your conclusion is obviously wrong, and an error message indicates bad MATLAB syntax.

Study your code below carefully. Determine the syntax and/or runtime error(s) and fix them.

Run the code with several inputs to test the improved version of **pythagorasForUnitCircle**:

```
function trueOrFalse = pythagorasForUnitCircle(angle)
% This function is written to verify the identity
%                  cos^2(angle)+sin^2(angle) = 1
% Input: angle... an angle in degrees
% Output: trueOrFalse ... 1 if identity is verified,
% 0 otherwise

    sincos = cos^2(angle)+sin^2(angle);
    trueOrFalse = (sincos=1);
```

Problem 4-B.14

You need to write a MATLAB function to compute $L(x)$ given by the following equation (where $0 < x < 1$):

$$L = 100\left(\frac{x}{0.6}\right)^{0.625}\left(\frac{1-x}{0.4}\right)^{-1.625}$$

The function you write as a first draft is the following:

```
function L = computeL(x)
% This function calculates L(x) per the problem
% spec
% Input: x
% Output: L

    L = 100 * x/0.6^0.625 * 1-x/0.4^-1.625;
```

Find and correct the errors in your first draft.

Chapter 5

Vector Operations

Let us take stock of where you should be in your understanding of MATLAB to this point. The high points include the topics listed below:

1. The purposes of the various MATLAB windows (Command, History, Current Directory, Editor, Workspace, Array Editor)

2. The concepts of MATLAB scalar variables and expressions

3. Concept and use of MATLAB scalar built-in functions

4. Specifics of MATLAB scalar operators (arithmetic, relational, logical)

5. Concept and use of the MATLAB current directory

6. Concept and use of MATLAB script files

7. Concept and use of MATLAB user-defined functions

8. Concepts and uses of MATLAB facilities for debugging scripts and user-defined functions

In a nutshell, you should know how to start MATLAB, how to use it to perform scalar operations, how to save your MATLAB work in scripts and user-defined functions, and how to debug your scripts and user-defined functions.

In this chapter, you will extend your knowledge to include MATLAB vectors. When we introduced scalars in Section 3-2, we contrasted scalars to vectors and emphasized that vectors in MATLAB are different in concept than vectors in mathematics. You must keep that distinction clearly in mind. When we use the term *vector* in this chapter, we mean a *MATLAB* vector. Scalars are simple numbers. MATLAB vectors are collections of numbers that are lined up. More precisely, MATLAB vectors are ordered, one-dimensional groups of scalars.

The goal of this chapter is to introduce you to vector operations in MATLAB. You have seen that MATLAB can be used as a sophisticated calculator for scalars. In the last chapter, you learned the ability to store MATLAB programs for later use or for sharing with others. But in a fundamental sense, with vectors, MATLAB capabilities go beyond what you have been able to do before with computational devices like calculators. Vectors are *groups* of scalar numbers.

There are two wins in being able to define and work with vectors in MATLAB. If you want to perform the same operation on each member of a vector, then you can perform all operations in one step. This makes describing a computation easier. The ability to group the numbers together allows you to tell MATLAB to perform the operation on all members of the group in *one* MATLAB command. The *second win* has to do with how you think about building solutions to technical problems. The ability to write MATLAB commands at a higher level (a single command to operate on a group of numbers all at once rather than being forced to write a separate command for each one of the scalars in the group) frees you to *think* at a higher level. You do not have to concentrate on writing commands for every scalar operation but, instead, can write one command that will cause MATLAB to apply a given operation to *all scalar elements* of an input vector.

The path you will follow in this chapter is the following:

- learn how to create vectors
- learn how to access vector elements
- learn about the two types of vectors (row and column vectors)
- extend your understanding of MATLAB built-in functions, operators, and expressions to include vectors

These bulleted items follow the same pattern you followed in Chapter 3 as you learned about MATLAB scalar operations, so most of the topics will be familiar. Now, you will extend what you already know to cover vectors.

5-1 Vector Creation

Start MATLAB and perform the following operation:

```
>> x = [1, 10]
```

You will see the following result:

```
x =
        1.00        10.00
```

You have a created a variable named **x**, and the value of **x** is a two-element vector. The first element of **x** is 1; the second is 10. The creation operator you have used to build **x** is the *square bracket operator*, **[...]**. Notice that each element of the vector you are creating is a simple scalar number.

The square bracket operator is quite versatile for creating vectors. This versatility extends to creating vectors that include other previously defined vectors. For example, once you have **x** defined as above, you can perform the following operation:

```
>> y = [-10, x]
```

From this, you will obtain the following:

```
y =
       -10.00        1.00        10.00
```

When you look casually at the assignment statement that creates **y** above, it seems to have two elements: −10 and **x**. But **x** has been defined as a vector with two elements. Hence, **y** ends up having three scalar elements. A useful MATLAB built-in function when you want to know the length of a vector is **length**:

```
>> lengthOfY = length(y)
lengthOfY =
        3.00
```

To create a vector, the square bracket operator need not include any scalar arguments at all. It can be used to "combine" two vectors. For example, if we wanted to *concatenate* **x** and **y** to a new vector **z**, we could use the following MATLAB command:

```
>> z = [x, y]
z =
        1.00    10.00    -10.00        1.00        10.00
```

The order of items is important when you use the square bracket operator; vectors are *ordered* groups of scalars.

The square bracket operator is the most flexible of the methods that MATLAB gives you for vector creation. You should use it when you want to create a vector in which the elements of the vector are unrelated to each other in some (usually linear) relationship or when you want to concatenate to link existing vectors to create a new vector.

In addition to the square bracket operator, two other common ways exist to create vectors: the built-in function **linspace** and the colon operator (`:`). Both are used when you want to produce a vector in which the spacing between vector elements is the same for the entire vector.

The **linspace** operator is best used when you know the first element and the last element of the vector you want to create, and how many elements you want your vector to contain. For example, if you want to create a vector **m** whose first element is 0 and whose last element is 10, and you want a total of five elements (scalars) in your vector, you could use the following command:

```
>> m = linspace(0, 10, 5)
```

You would get the following result:

```
m =

             0       2.50       5.00       7.50      10.00
```

The first argument to **linspace** is the first element of the vector you want to create, the second argument is the last element, and the third argument is the number of elements you want the constructed vector **m** to contain. Several relationships occur in a vector created by **linspace**:

$$numberOfIntervals = numberOfElements - 1$$
$$sizeOfInterval = \frac{lastElement - firstElement}{numberOfIntervals} \qquad \text{(EQ 5-1)}$$

When would these relationships be useful? (Hint: When you want to create a vector, do you always know the number of points you want your vector to have?)

The third argument to **linspace** (the number of elements you want in the vector) is optional. The number of elements *defaults* to 100 if you do not specify a third argument.[1] For example, suppose you create a vector **n**:

```
>> n = linspace(0,198)
```

The first element of **n** will be 0, the last element will be 198, one hundred
elements will be in **n**, and the space between each element will be the same. What
will that common interval size be? (Hint: Use the relationships given above in
Equation 5-1.) As you will see in the next chapter when you start learning about
creating data plots in MATLAB, **linspace** is quite useful, particularly the form
of **linspace** that allows the third argument to default to 100.

At this point you know how to create vectors in two ways: the square bracket
operator and the **linspace** built-in function. A third way to create vectors is to
use the colon operator (**:**). Like **linspace**, the colon operator produces vectors
whose elements are equally spaced. As with **linspace**, to use the colon
operator, you need to know the beginning and ending elements of the vector you
want to create. Unlike **linspace**, to use the colon operator you need to know
the interval size for the spacing between elements in the vector. For example, if
you know you want to create a vector **b** whose first element is 0, whose last
element is 8, and with an interval of 2 between each of the vector elements, you
can use the following:

```
>> b = 0:2:8
```

You will get the following result:

```
b =

          0      2.00      4.00      6.00      8.00
```

The first number appearing in the application of the colon operator is the value for
the starting element of the vector to be constructed. The second number (after the
first colon) is interval size. The third number (after the second colon) is the end
value. The action of the colon operator is to start at the beginning element. For
each succeeding element, add the value of the interval and continue doing that
until the end value is reached or exceeded.

When would you want to use the colon operator instead of **linspace**? That is,
when would it be easier to use the colon operator than the other two vector
creation methods? The relationships in Equation 5-1 allow you to use **linspace**
or the colon operator interchangeably, but when is the colon operator easier to
use?

---

1. A default value for an argument is one that MATLAB supplies if you do not specify it.

The last number appearing (of the three) in the application of the colon operator is almost the same as the second argument that you must give to **linspace**. But a difference is that the third number used in application of the colon operator is a limit, not the actual last value in the vector created. With **linspace** you are guaranteed that the last element of the vector created will be equal to the second argument you specify. With application of the colon operator, the guarantee is that the last element of the vector you create will be equal to *or smaller than* the last number in the application of the colon operator:

```
>> p = 0:2:9
```

This produces the following:

```
p =

          0       2.00      4.00      6.00      8.00
```

That is, **b** and **p** are identical vectors although the applications of the colon operator that produced them were different. Internally, to produce **b**, MATLAB sets the first element of the vector under construction to 0, adds 2 and sees that the sum is not larger than 8; therefore, it sets 2 as the second element of the vector being built, adds 2 again to get 4 and sees that 4 is not larger than 8. It sets 4 as the third element of the vector being built, et al., up to 8 when MATLAB sees that 8 is equal to the last number given to the colon operator. The result is the vector **[0, 2, 4, 6, 8]**.

On the other hand, when MATLAB is given the colon operator command to construct **p**, the operation of MATLAB is different. At the last step of the construction, MATLAB sees that 10 is larger than the limit set by the third number in the colon command, and hence does not change the vector being constructed at the last step. If you do not understand the phrase *MATLAB sees that 10 is larger than the limit* and, in particular, do not understand where the 10 came from, then review above how the colon operator is handled internally by MATLAB.

Just as **linspace** has a default form in which you can leave out the last argument and let MATLAB assume 100 equally spaced intervals, the colon operator has also a default form. You must, however, note which number (of the three in the full form of the colon operator) may be left out. If you leave out the *second* number of the three in the full form, then MATLAB will assume you want the interval between vector elements to be 1. The MATLAB command becomes the following:

```
>> q = 0.5:4
```

This results in the following answers:

```
q =
        0.50      1.50      2.50      3.50
```

The **q** vector was created to have a starting point of 0.5 (the first number given in the command above), not to exceed 4 (the last number given), and to have an interval of 1 between each element (the default value which MATLAB assumes if only two numbers are in an application of the colon operator).

Finally, you need to know how to use two special purpose vector creation built-in functions. The built-in function **ones(1,N)** creates a vector of length N whose elements are all the number ONE; the built-in function **zeros(1,N)** creates a vector whose elements are all the number ZERO. Both **ones** and **zeros** are built-in functions that take *two* arguments. For now, the first argument for ones and zeros must be 1. The second argument is the number of elements we want the created vector to have.[2]

For example, to create a vector with 3 elements, where each element is the number ONE, do the following:

```
>> vecAllOnes = ones(1,3)
vecAllOnes =
        1.00      1.00      1.00
```

As a second example, create a vector with four elements in which each element is the number ZERO:

```
>> vecAllZeros = zeros(1,4)
vecAllZeros =
        0         0         0         0
```

---

2. Right now, the discussion is focusing on one type of vector: row vectors. When we discuss column vectors and arrays, we will return to the meaning of the arguments for **ones** and **zeros**.

1. Creation of vectors may be accomplished by the square bracket operator, the `linspace` built-in operator, or the colon operator.

2. Use the square bracket operator to construct an arbitrary vector.

3. Use `linspace` when you want an equally spaced vector, and you know the starting and ending elements and the number of elements you desire.

4. Use the colon operator when you want an equally spaced vector, and you know the starting value, ending value limit, and the interval size you desire.

5. To create a vector with N elements where each element is the number ONE, use ones (1,N).

6. To create a vector with N elements where each element is the number ZERO, use zeros (1,N).

Synopsis for Section 5-1

5-2   Accessing Vector Elements

If you look back to Section 3-2, Section 3-3, and Section 3-4, where you learned about scalar computations in MATLAB, you will see a progression:

- Scalar variable creation
- Fetching and setting scalars variables
- Built-in scalar functions
- Scalar operators
- Scalar expressions

We are following the same general progression here to discuss vectors and operations on vectors in MATLAB. In the last section you learned how to create vectors. In this section you will learn about vector access or how to fetch and set vectors.

Unlike scalar access, vector access has two flavors:

1. fetching or setting an entire vector
2. fetching or setting one or more elements of a vector

The first, fetching or setting an entire vector, is similar to scalar access. You create a variable to be a vector variable or reset the value of a vector variable by using an assignment command; you did this in the last section. You can fetch the value of a vector variable for use in other commands by typing the name of the variable. Once again, you did that in the last section.

5-2.1   Fetching Elements of Vectors

But unlike scalars, vectors have *parts*. In MATLAB, you can access one or more elements of a defined vector through an operation called *indexing*. Recall that vectors are *ordered* groups of scalars. In the vector **[7, 3, 4, 2]**, there is exactly one first element (the scalar 7), exactly one second element (the scalar 3), and so on. MATLAB allows you to dig out specific elements in a vector. Put in MATLAB terms, MATLAB allows you to *index* specific elements.

In the Command Window, execute the following:

```
>> x = [7, 3, 4, 2];
```

Index the first element of **x**, and set the variable **firstElement** to the first element of **x**:

```
>> firstElement = x(1)
firstElement =
          7.00
```

You can use an indexed value from a vector the same way you would a variable. For example, set the variable **y** to the value of the third element of **x** plus 2:

```
>> y = x(3) + 2
y =
            6.00
```

Fetching the value of one element of a vector is just the beginning. Suppose you want to create a new vector **y** that has two elements: the third and fourth elements of **x**. The following commands in MATLAB will accomplish that objective:

```
>> ndx = [3, 4];
>> y = x(ndx);
```

In the first command, a vector **ndx** is created, and in the second, **ndx** is used as the *index* into **y**. You could have accomplished the same thing in one command:

```
>> y = x([3,4]);
```

The first form in which the **ndx** vector is explicitly named and set is preferable when first learning vector operations.

Let us go a step further and ask how you can extract the odd-numbered elements of **x** to a vector **z**. Given that we know that **x** has four elements, we could use the following:

```
>> ndx2 = [1,3];
>> z = x(ndx2);
```

A more general way exists, a way that will work no matter how many elements **x** contains. The key to understanding this other solution is to think back to the three methods for creating vectors and the situations for using each. The path here is to select the vector creation method whose *use criteria* best matches what we want to do.

What we want to create is an **index vector** for all the items in **x** we want to extract, i.e., the odd-numbered elements. We know the following about the index vector we want to construct:

1. the starting element value: 1

2. the interval between the element values: 2

3. the end element value: **length(x)**

Putting all this together, we can extract the odd elements of **x** with the following commands:

```
>> x = [7, 3, 4, 2];
>> ndxOddElements = 1:2:length(x)
ndxOddElements =
            1.00      3.00
>> z = x(ndxOddElements)
z =
            7.00      4.00
```

The above commands will extract the odd elements from x no matter how many elements **x** contains:

```
>> x = [7, 3, 4, 2, 35, 23, 11];
>> ndxOddElements = 1:2:length(x)
ndxOddElements =
            1.00      3.00      5.00      7.00
>> z = x(ndxOddElements)
z =
            7.00      4.00     35.00     11.00
```

The lesson to be learned from this last example is this: You should always construct solutions, whether with MATLAB or with any other tools, to develop general solutions if possible. In the example for extracting the odd-numbered elements from a vector, if you know the specific vector you are working with, then a quick and dirty solution is the easy way. If you do not have a specific vector and want your solution to work with an arbitrary vector, then you have to think some more. Working toward general solutions is good practice but typically amounts to a trade-off between generality and difficulty: The more general your solution the harder it can be to develop.

There is one more item to cover in this section on fetching elements of vectors. MATLAB gives you the means to easily index the last element of a vector. The word **end** used in an indexing expression tells MATLAB that you are referring to the last element in a vector:

```
>> x = [3, 8, 1, 9];
>> y = x(end)
y =
            9.00
```

The above tells MATLAB to fetch the last element of **x**. **Note the following carefully: you cannot use end except** *directly* **in an indexing expression.** Above in the cases where we set up an indexing vector, then used it to access specified elements of a vector, we could not have made use of the MATLAB indexing **end**.

Suppose we want to access the next to last element in a vector. In that case, we could use the following:

```
>> x = [3, 8, 1, 9];
>> y = x(end-1)
y =
          1.00
```

As a final example, suppose we want to access the last *two* elements of a vector:

```
>> x = [3, 8, 1, 9];
>> y = x(end-1:end)
y =
          1.00      9.00
```

The MATLAB indexing **end** is shorthand for getting the length of a vector and using it for an index back into the vector:

```
>> areLengthAndEndTheSame = x(length(x)) == x(end)
areLengthAndEndTheSame =
          1.00
```

5-2.2   Setting Elements of Vectors

So far, you have seen one side of vector access: fetching values. The other side, setting values, will now be easy to understand based on the previous discussion of vector fetching. Though fetching operations always appear on the right side of an assignment operator, the same forms of indexing work to set elements of a vector if the indexed vector appears on the *left* side of an assignment operator. Suppose we have this initial value for vector **y**:

```
>> y = [33,10,105,57];
```

We want to reset the second element of **y** to be −1. The following set of commands accomplishes this goal:

```
>> ndx= 2;
>> y(ndx) = -1
y =
          33.00    -1.00   105.00    57.00
```

You could use as a shortcut:

```
>> y(2) = -1
y =
        33.00     -1.00   105.00     57.00
```

Let us also change the last element of **y** to be 0:

```
>> y(end) = 0
y =
        33.00     -1.00   105.00        0
```

Setting a single element in a vector is easy. What you are doing is telling MATLAB what element of the vector you want to reset (by indexing to it), and using an assignment statement to reset that element to the new value you specify.

Conceptually, what you just did and resetting more than one vector element are the same. The only difference is you are now indexing to multiple elements *and the number of elements you specify in the index must be equal to the number of elements that you give on the right hand side of the assignment.* Suppose you start with a vector **b**:

```
>> b = [4, 8, 2, 1];
```

You want to reset the first element to −22 and the last element to −100. The commands below will accomplish the task:

```
>> ndx = [1, length(b)];
>> b(ndx) = [-22, -100]
b =
        -22.00      8.00      2.00  -100.00
```

Here is an alternative way of doing the above:

```
>> b = [4, 8, 2, 1];
>> replacementFirstAndLast = [-22, -100];
>> b([1,end]) = replacementFirstAndLast
b =
        -22.00      8.00      2.00  -100.00
```

A variation on the second form above would be what you would want if your task were to write a function taking as arguments (a) the vector in which the replacements were to be made and (b) a vector containing the replacement values for the first and last elements. Make sure MATLAB is open and write a function that: (1) takes as input arguments vector **b** and a replacement vector **replaceV** (2 elements) for the first and last values, and (2) returns the modified vector after replacement.

Here is a general rule for vector element replacement using the assignment operator: The number of elements you are replacing (on the left-hand side of the assignment operator) must be equal to the number of elements on the right side of the assignment operator. For example, given a ten element vector **w**, the following operation is legal:

```
>> w = 1:10;
>> w([1,7]) = [45, 22];
```

The following operation is **not** legal:

```
>> w = 1:10;
>> w([1,4,5]) = [102, 211];
```

An exception to the rule described above exists. Suppose you want to reset a number of elements in a vector to the same scalar value:

```
>> w = 1:5;
>> w([3,2]) = [99, 99]
w =
          1.00     99.00     99.00      4.00      5.00
```

But MATLAB provides a short cut way to accomplish the same result:

```
>> w = 1:5;
>> w([3,2]) = 99
w =
          1.00     99.00     99.00      4.00      5.00
```

If the left-hand side of an assignment operator is a vector access and the right-hand side is a single scalar, then all elements indicated on the left-hand set are set to the single scalar value.

Suppose your purpose is not to reset an element, but to erase it. For example, suppose vector **a** is **[4, 7, 2, 1, 33]** and you want to erase the third element. Here, erase means you want to *remove* the third element from **a**, leaving **a** with four elements. The command to remove the third element from **a** is the second command below:

```
>> a = [4, 7, 2, 1, 33];
>> a(3) = []
a =
          4.00      7.00      1.00     33.00
```

The square bracket operator was one method for vector creation. The square bracket operator with nothing inside the bracket is saying "create an empty vector." The assignment of the empty vector to the third element of **a** has the effect of erasing that third element.

That's it! There is no more you need to know at this point for accessing vector elements. As a final example, suppose you have vector **z**:

```
>> z = [44, 2, 9, 77, 66, 32, 99];
```

You want to set every third element to 0, starting with the third element. The following function will accomplish that task:

```
function originalVector = setEveryThirdZero(originalVector)
 % changes every third element of the input vector to zero,
 % and returns the modified vector
 % INPUT: originalVector
 % OUTPUT: originalVector modified so that
 %                            every third element is 0

   ndx = 3:3:length(originalVector);
   originalVector(ndx) = 0;
```

ndx is an index vector starting at index 3 and continuing with every third *index*. The last command performs the operation of zero setting once we have the index vector we need.

This last example is important to you because if you understand it immediately, it means that you understand these concepts. If you do not understand the example, then you should start MATLAB, try the example, but remove the semicolons so you can see the results of each command. If you still do not understand the example, go back and review this section.

1. Access to a complete vector is similar to scalar access.

2. Accessing an element(s) in a vector depends on the use of an *index* into the vector.

3. To delete an element(s) in a vector, the empty square bracket is used.

4. To find the length (the number elements) of a vector V, use the `length` built-in function: `length(V)`.

5. When setting elements of a vector, the number of elements being set (indexed on the left-hand side of an assignment operation) must be equal to the number of elements in a vector on the right-hand side of the assignment operation. The exception is that a scalar on the right-hand side can be used to set multiple vector elements.

Synopsis for Section 5-2

5-3   Row Vectors and Column Vectors, and the Transpose Operator

In Section 5-1 and Section 5-2, you learned how to create and access vectors. Though we have treated all vectors as similar, two major vector types exist: row vectors and column vectors. You need to understand the difference between row and column vectors to understand parts of Section 5-4 below on vector functions, operators, and expressions.

For now, the easiest way to distinguish between row and column vectors is their display method. As the name implies, a row vector displays horizontally in MATLAB. All the material in Section 5-1 and Section 5-2 was about row vectors.

Column vectors display vertically. Everything we have said in Section 5-1 and Section 5-2 about row vectors is true for column vectors *except for one item*, which is in the use of the square bracket operator to create a vector. To create a row vector with the square bracket operator, you place into the square brackets the elements you want to make part of the vector *separated by commas or spaces*.[3]

```
>> rowVector_X = [88, 24, 35]
rowVector_X =
          88.00     24.00     35.00
```

To create a column vector with the same three elements, you use the square bracket operator, and inside the square brackets, put the elements you want to make part of the vector *separated by semicolons*:

```
>> columnVector_X = [88; 24; 35]
columnVector_X =
          88.00
          24.00
          35.00
```

Everything in Section 5-1 and Section 5-2 applies equally to row and column vectors except for this one point about creation of vectors with the square bracket operator.

Converting a row vector to a column vector, or vice versa, is easy because MATLAB gives you a built-in operator to accomplish the conversion task: the **transpose**

---

3. If the elements are separated by a space, a row vector will be produced, too. In this book, we will stick to using a comma whenever we want to build a row vector with the square brackets operator.

operator. The transpose operator is invoked by placing a single quote mark (') after the vector (or vector expression) you want to convert:

```
>> z = 1:3:11
z =
          1.00      4.00      7.00     10.00
>> zAsColumn = z'
zAsColumn =
          1.00
          4.00
          7.00
         10.00
```

Or you can convert back to a row vector:

```
>> >> zBackToRow = zAsColumn'
zBackToRow =
          1.00      4.00      7.00     10.00
```

A good way to think about the transpose operator is that it flips its argument. For vectors, this amounts to changing a row vector to a column vector or to changing a column vector to a row vector. The elements of the vector remain the same; it is a matter of whether the vector is a row vector or a column vector.

How can you make direct use of the other two vector creation methods (**linspace** or the colon operator) to make a column vector? You can't. If you want to build a column operator and find it easiest to use **linspace** or the colon operator, build a row vector first, and then convert it to a column vector using the transpose operator.

At this point, you are probably scratching your head and asking why you need row and column vectors if they amount to the same thing. You will understand the difference between row and column vectors after you learn about arrays in the next chapter. Think of it this way: An array is like a table of data, like an Excel table. There are rows you distinguish from left to right in such a table and there are columns you distinguish from top to bottom. A row vector is a *horizontal slice* through such data tables (arrays), and a column vector is one *vertical slice*. Getting used to the idea of vectors coming in the two flavors (row and column) will help for later discussion.

1. A row vector is a horizontal vector.

2. A column vector is a vertical vector.

3. All operations in Section 5-1 and Section 5-2 are true for row and column vectors except the creation of a column vector using the square bracket operator utilizes a semicolon, and the creation of a row vector utilizes a comma.

4. A column vector cannot be created directly using the colon operator or the `linspace` built-in function.

5. The built-in transpose operator (`'`) can be used to convert a row vector to a column vector, or vice versa.

Synopsis for Section 5-3

5-4   Vector Built-in Functions, Operators, and Expressions

As we did for scalars in Chapter 3, you have learned how to fetch vectors and elements in vectors and how to set vectors and elements in vectors. The next step is to build commands using vectors. We will follow the same basic path for vectors that we followed before when you learned about scalars. A good deal of what you learned for scalars in Section 3-4 will be the starting point for discussion here.

5-4.1   Vector Built-in Functions

The good news is that most MATLAB built-in functions that take one argument work similarly whether they are applied to a scalar or to a vector. There is no bad news.

Suppose you have a group of angles for which you want to find the sine. You could apply the built-in function **sin** to each scalar angle in the group. MATLAB provides an easier way. First, create a vector with all the angles you want to use; then, apply **sin** to the vector. Remembering that angles in MATLAB are expressed in radians, you can do the following:

```
>> angleSet = [0, pi/2, pi, 3*pi/2, 2*pi];
>> sinOfAngles = sin(angleSet)
sinOfAngles =
          0      1.00      0.00     -1.00     -0.00
```

The **sin** built-in function is applied to each of the elements of **angleSet**. The result is a vector with the same length as **angleSet** in which each element is the result of **sin** applied to the corresponding element in **angleSet**.

If a built-in function of one argument, like **sin**, is given a row vector as its input argument, then the output result will be a row vector. If it is given that the input is a column vector, the output will be a column vector. For example, redo the example above and give the **sin** function a column vector of angles:

```
>> angleSet = [0; pi/2; pi; 3*pi/2; 2*pi];
>> sinOfAngles = sin(angleSet)
sinOfAngles =
         0
      1.00
      0.00
     -1.00
     -0.00
```

In addition to the built-in functions like **sin** that process each element of a vector, some vector operators act on an entire vector at once. Table 5-1 shows a few of this type of built-in vector functions. You will experience the usefulness of these functions as you work the chapter exercises. Other built-in vector functions perform *cumulative computation* on all members of a vector argument.

**Table 5-1: Sample of Built-In Vector Functions
that Act on Entire Vectors**

vector function	meaning	example input	result
sum	compute the sum of all vector elements	**sum([1,3,2])**	6
max	find the maximum of all vector elements	**max([1,3,2])**	3
min	find the minimum of all vector elements	**min([1,3,2])**	1
mean	find the arithmetic mean of all vector elements	**mean([1,3,2])**	2
sort	sort the elements of a vector in ascending order	sort([3,1,2])	[1, 2, 3]

5-4.2   Vector Operators

There is good news and bad news. The good news is that the set of scalar operators are *almost* the same as the set of vector operators. The bad news is that you have to learn the differences and the generalizations of the scalar operators compared to vector operators.

We will use four classes of vector operators, as was the case for scalar operators discussed in Section 3-4.1.1.

1. **Special operators:** This class of operators includes the assignment operator, the transpose operator, the colon operator, and the square bracket operator as covered earlier in this chapter.

2. **Arithmetic operators:** This class of operators enables normal operations of arithmetic for two scalars, two vectors, or a vector and a scalar.

3. **Relational operators:** This class of operators enables comparing two scalars, two vectors, or one scalar and one vector.

4. **Logic operators:** This class of operators enables combining the results of one or more relational tests.

There are not many changes from the original list of operators for scalars given in Section 3-4.1.1 from the list above. Class 1 now includes the assignment operator, the colon operator, and the transpose operator. Class 2 and Class 3 includes vectors and combinations of a scalar and a vector. Class 4 is unchanged. As in Section 3-4.1.1, we will look at Class 2, 3, and 4 operators in turn.

Table 5-2: Scalar and Vector Arithmetic Operator Class (cell-by-cell)

Operator name	Operator Symbol	Example
unary plus	**+**	$+s$ $+\bar{\mathbf{x}}$
unary minus	**–**	$-s$ $-\bar{\mathbf{x}}$
addition	**+**	$s+t$ $\bar{\mathbf{x}}+\bar{\mathbf{y}}$ $\bar{\mathbf{x}}+s$
subtraction	**–**	$s-t$ $\bar{\mathbf{x}}-\bar{\mathbf{y}}$ $\bar{\mathbf{x}}-s$ $s-\bar{\mathbf{x}}$
multiplication: cell-by-cell	**.***	$s.*t$ $\bar{\mathbf{x}}.*\bar{\mathbf{y}}$ $s.*\bar{\mathbf{x}}$ $\bar{\mathbf{x}}.*s$
right division: cell-by-cell	**./**	$s./t$ $\bar{\mathbf{x}}./\bar{\mathbf{y}}$ $\bar{\mathbf{x}}./s$ $s./\bar{\mathbf{x}}$
exponentiation: cell-by-cell	**.^**	$s.^{\wedge}t$ $\bar{\mathbf{x}}.^{\wedge}\bar{\mathbf{y}}$ $\bar{\mathbf{x}}.^{\wedge}s$ $s.^{\wedge}\bar{\mathbf{x}}$

5-4.2.1 Vector Arithmetic Operators (cell-by-cell operators only)

Table 5-2 updates Table 3-4 where we summarized arithmetic scalar operators. In the table, $\bar{x}$ and $\bar{y}$ are vectors; s and t are scalars. The updated version includes both scalars *and* vectors.

We need to understand *cell-by-cell operations* in the abstract before we turn to detailed discussion of the operators in Table 5-2. The meaning of this phrase is most easily understood by example. Suppose you have a vector **x** and a vector **y**:

```
>> x = [1, 10, 4];
>> y = [3, 2, 9];
```

Examine a cell-by-cell operation between **x** and **y**:

```
>> z = x + y
```

Using addition means that the same operation is carried out for each pair of *corresponding elements* in **x** and **y**. The first element of **z** will be the first element of **x** plus the first element of **y** for a total of 4. Similarly the second element of **z** will be the second element of **x** plus the second element of **y**: 12. And the third element of **z** will be 13.

```
>> z = x + y
z =
            4.00         12.00         13.00
```

Two rules must be true to enable any cell-by-cell operation between two vectors **x** and **y**, that is for all of the entries in Table 5-2 except for unary plus and unary minus where there is only one vector to worry about.

1. Both **x** and **y** must be a row vector, Or **x** and **y** must be a column vector.
2. The number of elements in **x** and **y** must be the same.

If you think about the abstract idea of cell-by-cell operation (the same operation is carried out on corresponding vector elements) you will understand why these two rules are necessary: (1) the two vectors must be of the same type (row or column) to have corresponding elements, and (2) the vectors must be of the same length or else some elements in the longer vector will not have corresponding elements in the shorter. In a later chapter, we will extend the table of arithmetic operators one last time and go beyond cell-by-cell operations.

Although Table 5-2 looks complicated, it is not. The first four rows of Table 5-2 use the common operator symbols that you learned for scalar operators. Suppose you have a vector **x**:

```
>> x = [-1, 10, -22];
```

Suppose you want to flip the sign of each of the elements of **x**. This is easily accomplished with the unary minus operator:

```
>> negatedX = - x
negatedX =
             1.00              -10.00              22.00
```

This is a cell-by-cell operation because the unary minus (scalar) operator is applied to each of elements of **x**. The unary plus and the unary minus operator work in this fashion. Unary plus and unary minus work for scalars and vectors. What we are doing in Table 5-2 is extending the unary minus and unary plus to scalars and vectors. So, the argument to the unary operators is either a scalar or a vector. For unary minus, the scalar argument is flipped in sign; each element of the vector argument is flipped in sign.

How is this possible? Unary minus means two different things depending on what the argument is given to the operator, whether scalar or vector. The answer to this question is easy, yet profound from a computer science viewpoint. MATLAB internally has multiple versions of unary minus. When MATLAB is asked to compute an expression that contains a unary minus operator, MATLAB determines what to do based on the argument. For cell-by-cell operations, this is conceptually easy to understand. The innards of MATLAB handle all the mechanics of the decision for which unary minus applies, and the innards hide all of that decision from you by using a methodology called *object-oriented-programming* and more specifically by using a technique called *operator overloading*. We will not concern ourselves in this book further with how operator overloading works internally in MATLAB, but you must recognize that operator overloading is part of MATLAB, and for each specific case, like the two cases for the unary operators, you must understand how MATLAB handles each.

The third and fourth rows of Table 5-2 take a little more care to understand. First, the minus (–) and the plus (+) operators work for *four* cases: (a) both arguments are scalars, (b) both arguments are vectors, (c) the first argument is a scalar and the second a vector, and (d) the first argument is a vector and the second a scalar. Case (a) is your old friend, the scalar plus operator:

```
>> s = 3 + 2
s =
             5.00
```

Case (b) is a cell-by-cell operation: Corresponding elements in the two vector arguments are added to produce the result:

```
>> x = [2, 3];
>> y = [22, 45];
>> z = x + y
z =
        24.00              48.00
```

Cases (c) and (d) are the mixed cases in which one argument is a vector and the other a scalar. The easiest way to understand (and remember) how cases (c) and (d) work is to think of them as two-step computations. First, the scalar is converted into a vector so the type of the converted vector (row or column vector) is the same as the other argument, the length of the converted vector is the same as the other argument, and each element of the converted vector is equal to the scalar value. Second, the operation is performed as a standard cell-by-cell operations. Examine this operation:

```
>> x = [3, 9, 7];
>> s = 10;
>> z = s + x
z =
        13.00            19.00            17.00
```

The operation to compute **z** is easy to understand if you think of **s** as converted to **s2** = **[10, 10, 10]** first. Then, perform **s2 + x** as a standard cell-by-cell operation.

Rows 5, 6, and 7 (the last three rows) of Table 5-2 involve a new element: the Dot (**.**) symbol. Its three uses are: **.*** meaning cell-by-cell multiplication, **./** meaning cell-by-cell *right* division, and **.^** meaning cell-by-cell exponentiation. When MATLAB sees a command with a Dot where it is expecting an operator, it looks at the character immediately following. For example, you should think of **.*** as one operator, the cell-by-cell multiplication operator, because that is the way MATLAB understands **.***.

Why the change to using the Dot for multiplication, division, and exponentiation? You will have to defer getting the full answer to that question until the later chapter on array and matrix operations. The answer has something to do with there being more than one kind of multiplication, division, and exponentiation that we have to treat differently at a conceptual level.

As the case for cell-by-cell addition and subtraction, cell-by-cell multiplication, division, and exponentiation have four cases that you need to understand: (a) both arguments are scalars, (b) both arguments are vectors, (c) the first argument is a scalar and the second a vector, and (d) the first argument is a vector and the second a scalar.

For cell-by-cell multiplication, case (a) is the familiar scalar multiplication except that we are using .* for the operator. Case b is cell-by-cell multiplication applied to two vectors:

```
>> x = [2; 3];
>> y = [20; 30];
>> z = x .* y
z =
            40.00
            90.00
```

Cases (c) and (d) are analogous to the situation for cell-by-cell addition and subtraction with mixed arguments (one scalar and one vector). A way to understand mixed argument cell-by-cell operations is to think of the scalar as being converted to a vector with all elements equal to the value of the scalar. For example:

```
>> x = [3, 6, 7];
>> s = 11;
>> z = s .* x
z =
            33.00            66.00            77.00
```

Note that z2 = x .* s would have produced the same result.

Cell-by-cell division of vectors and scalars has two forms: *right* division and *left* division. In this chapter, we will cover right division. The symbol for cell-by-cell vector right division in MATLAB is ./. To understand the entry in Table 5-2 for cell-by-cell vector right division, all you need to do is play back the same story we had for cell-by-cell vector multiplication. Right division, when both arguments are scalars, is the standard arithmetic division operation. Right division, when both arguments are vectors, is straightforward cell-by-cell division. Finally, right division of mixed scalar/vector arguments amounts to first converting the scalar to the appropriate vector and then performing vector cell-by-cell division.

Table 5-2 has one more vector operator: exponentiation. The symbol for cell-by-cell vector exponentiation in MATLAB is .^. And the story for the four cases shown in the last row of Table 5-2 (exponentiation) is similar to the cell-by-cell vector multiplication operation and the cell-by-cell right division operation.

 You should not underestimate the power of the ability to perform the same arithmetic operation on an entire set of scalar numbers (vectors). You have taken the first step toward learning the MATLAB mind-set for problem solving. The central feature of this mind set is that it enables you to think about the problems you want to solve at a level that is appropriate to the problem itself. To illustrate, consider the following problem:

You are asked to calculate the weekly take home pay for a group of employees at your company. Four employees are in the set; their employee ID numbers are 87, 43, 99, and 5. The employees in

Table 5-3: Data Table for Example

employee ID	regular hrs	overtime hrs
87	40	3
43	40	8
99	32	0
5	40	15

this group have a wage rate of $35 per hour, and $65 per hour when they work overtime. Table 5-3 lists the hours worked by the employees. You decide to solve the problem by writing a MATLAB function.

The first thing to do is to ask yourself what your function will take as inputs and what it will provide as outputs. The inputs to the problem are (a) a list of the employee IDs, (b) a list of the number of regular pay hours each employee worked, (c) a list of the number of overtime pay hours each employee worked, (d) the regular pay rate, and (e) the overtime pay rate. You can express the first three inputs as vectors:

```
>> employeeID = [87; 43; 99; 5];      % hrs
>> regularHrs = [40; 40; 32; 40];     % hrs
>> overtimeHrs = [3; 8; 0; 15];       % hrs
>> regularPayRate = 35;               % dollars/hr
>> overtimePayRate = 65;              % dollars/hr
```

The values for each of the three vectors (**emoployeeID, regularHrs, overtimeHrs**) is one of the columns of problem data table, Table 5-3. Hence we have represented each vector as a column vector (instead of a row vector) to have the closest match between the problem statement and the representation of it that we give to MATLAB. These setup operations to define specific values for the problem should be done in the Command Window, and these inputs will be passed to the function you are going to write.

The outputs from your function must include the gross paycheck amount for each employee. But as a safeguard, you should output the employee ID vector. The idea is that your function will output a vector containing employee ID numbers, and a second vector in which corresponding elements to the ID number vector will be the gross pay for that employee.

The computation to solve the problem is simple provided you focus first on what you would do to answer the question for one of the employees. For example, to compute the gross take home pay for employee #87, we could multiply the regular hours worked by #87 times the regular hourly rate, and add that value to the product of the hours of overtime worked by #87 and the overtime hourly rate. Knowing how to perform the computation for one employee, we can quickly write the MATLAB instructions for performing the computation for all employees by representing the data as appropriate vectors and performing the calculation as cell-by-cell computations:

```
>> regularGrossPay = regularPayRate .* regularHrs;      % dollars
>> overtimeGrossPay = overtimePayRate .* overtimeHrs;   % dollars
>> totalGrossPay = regularGrossPay + overtimeGrossPay;  % dollars
```

These computations will be in the body of the function you write. The names of the variables may not be identical to the names of the variables here, so you will have to make the variable names consistent with the function definition you write.

As an exercise, write a MATLAB function that brings all this together. Name your function **calculatePay**. To make the exercise explicit, make the first line of **calculatePay** as follows:

```
function [empID, totGrossPay] =
calculatePay(empID,regPay,otPay,regHrs,otHrs)
```

Test your function with the values for the input variables given in the problem. Determine what the answer should be for, say, employee #87. Run your function and verify that the actual value output from the function meets your expectation.

Now, how would you have to change what you did if you were dealing with four thousand employees? You would have to get the expanded data into MATLAB. Suppose you can import the data in from a supplied computer file. What must you change in your function **calculatePay**? The answer is, nothing. Verify for yourself that this is correct, and then think about why this is such an important point.

5-4.2.2   Vector Relational Operators

The relational operators that you saw in Table 3-4 that were discussed in the context of scalar operators are the same as the operators for cell-by-cell vector relational operators. The purpose of relational operators is to enable you to make numerical comparisons. Earlier, you learned about making comparisons between scalars. Now, you will extend your understanding to allow you to make comparisons between two vectors or between one scalar and one vector.

Table 5-4: Cell-by-Cell Relational Operator Class

Operator Name	Operator Symbol	Example
LESS THAN	<	x < y
LESS THAN, EQUAL TO	<=	x <= y
EQUAL TO	==	x == y
NOT EQUAL TO	~=	x ~= y
GREATER THAN, EQUAL TO	>=	x >= y
GREATER THAN	>	x > y

In Table 5-2, each case of using the operator for scalars and vectors was listed. In Table 5-4, the possible mixture of scalars and vectors is summarized by allowing **x** and **y** (in the examples column of Table 5-4) each to be a scalar or a vector. The bottom line is the same as before: There are four cases for each row of Table 5-4: **x** and **y** are scalars, **x** and **y** are vectors, **x** is a scalar and **y** is a vector, and **x** is a vector and **y** is a scalar. You learned about the first case (both arguments are scalars) in Section 3-4.1.1.2. The case in which both arguments are vectors is cell-by-cell comparison. The mixed cases are understood most easily by thinking of converting the scalar to an appropriate vector. All this should be familiar to you.

Suppose you have two vectors **x** and **y**, and you want to find out which elements of **x** are greater than the corresponding elements of **y**. The following code solves this comparison operation:

```
>> x = [4, 2, 9, 3];
>> y = [3, 3, 10, 1];
>> z = x > y
z =
          1.00            0            0          1.00
```

In compliance with the two general rules for cell-by-cell operations, **x** and **y** are of the same type (row vector), and both have the same length (four elements). The result of the cell-by-cell comparison of vectors **x** and **y** is a vector of the same type (**z** is a row vector) and with the same length (**z** has four elements) as the two argument vectors. The resultant vector (**z**) is made up of elements that are 1 or 0, where an element at some location n in **z** of 1 means the relational test is true for **x(n) > y(n)**, and conversely for the meaning of a 0 at some location *m*. For example, if *n* is 1, the value `z(1)` is True (the logic value is 1) because `x(1) > y(1)`.

Suppose you want to extend the earlier example given in the last section. You have written and tested the function `calculatePay`. Extend `calculatePay` to include another vector output that answers this question: "Which employees will get gross pay greater than $1,500?"

5-4.2.3   Vector Logic Operators (Cell-by-Cell) and Built-in Logic Functions (Accumulate)

In this section, you will learn about vector cell-by-cell logic operators and a set of built-in logic functions that allow you accumulate a group of relational tests into one answer.[4]

Table 5-5: Logic Operator Class (Cell-by-Cell)

Operator Name	Operator Symbol	Example
combine AND	**&**	x & y
combine OR	**|**	x | y
flip NOT	**~**	~ x

First, we will focus on the same operators you first saw in Table 3-5. The update to the cell-by-cell logic operators that can be applied to scalars, vectors, or mixed scalar/vector arguments is shown in Table 5-5.

As earlier, **x** and **y** in Table 5-5 can each be a scalar or a vector. Thus, the first two rows in Table 5-5 each stand for four cases as before. Moreover, understanding the entries in Table 5-5 is easy if you remember that the computations performed by these operators are cell-by-cell, and that for mixed scalar/vector arguments you can think of converting the scalar to an appropriate vector first, and then performing the computation.

Suppose you want to answer some questions about two stocks you are thinking of buying for investments. Recent prices for the two stocks can be represented in vectors with the price for the earliest date first and the price for the most recent date last in the vector:

---

4. Logical built-in functions for vectors are placed in this section rather than in the general section on vector built-in functions (Section 5-4.1) because, conceptually, they fit better with the logic operators.

```
>> priceStock_1 = [25.11, 24.27, 24.14, 23.99];⁵
>> priceStock_2 = [80.97, 79.69, 79.75, 80.69];⁶
```

"When was the price of stock #1 above $25 *and* the price of stock #2 above $80?" Two relational tests are involved in this question, each to be answered through application of cell-by-cell relational tests:

```
>> stock1Above25 = priceStock_1 > 25
stock1Above25 =
        1.00              0              0              0
```

```
>> stock2Above80 = priceStock_2 > 80
stock2Above80 =
        1.00              0              0           1.00
```

Following the two relational tests, you can answer the And question by applying a logical combination of the results from the two relational tests:

```
>> stock1Above25_AND_stock2Above80 = stock1Above25 &
stock2Above80
stock1Above25_AND_stock2Above80 =
        1.00              0              0              0
```

Likewise, you can ask "When was the price of stock #1 above $25 *or* the price of stock #2 above $80. The *logical or* means either one or the other, or both:

```
>> stock1Above25_OR_stock2Above80 = stock1Above25 |
stock2Above80
stock1Above25_OR_stock2Above80 =
        1.00              0              0           1.00
```

As a final example, the following answers the question "When was the price of stock #1 above $25 *or* the price of stock #2 **not** above $80?

```
>> stock1Above25_OR_stock2NotAbove80 = stock1Above25 |
~stock2Above80
stock1Above25_OR_stock2NotAbove80 =
        1.00           1.00           1.00              0
```

A good self-test you should use to see if you understand the vector cell-by-cell logic operators is to verify the last result by hand.

---

5. These are the closing stock prices for INTEL for August 4-7, 2003.
6. These are the closing stock prices for IBM for August 4-7, 2003.

Having these basic cell-by-cell logic operators in hand, we can go on to the built-in vector logic functions. Although MATLAB contains more built-in vector logic functions than the three shown in Table 5-6, we will cover these three only.

Table 5-6: Selected Vector Built-in Logic Functions

function name	what the function does	example
all	tests if all elements of a vector input are True (non-zero)	all(x)
any	tests if any element of a vector input is True (non-zero)	any(x)
find	finds **indices** for non-zero elements of an input vector	find(x > 22)

The first focus will be the first two entries in Table 5-6: the vector built-in functions **all** and **any**. Using **any**, you can answer the question "Was stock #1 over $25 for *any* day?"

```
>> stock1Over25ForAnyDay = any(priceStock_1 > 25)
stock1Over25ForAnyDay =
          1.00
```

Similarly, you can use **all** to answer the question "Was the stock price for stock #1 over $25 for *all* days?"

```
>> stock1Over25ForAllDays = all(priceStock_1 > 25)
stock1Over25ForAllDays =
          0
```

Like the accumulating numerical built-in functions **sum** and **mean**, the **any** and **all** built-in functions accumulate a result. And like **sum** and **mean**, built-in logic functions like **any** and **all** can be powerful tools. The built-in function **find** is more powerful; **find** returns the index or indices where relational tests are True within a target vector. You must understand and master **find** to use it in the example below.

Suppose you want to answer the following question: What was the price of stock #1 on those days when stock #2 was over $80? Let us start with the same price sequences we used earlier in this section:

```
>> priceStock_1 = [25.11, 24.27, 24.14, 23.99];
>> priceStock_2 = [80.97, 79.69, 79.75, 80.69];
```

Answer the question manually by observing the two price vectors above. Be methodical in noting the process you follow to answer the question.

What you probably did to answer the question was to look at **priceStock_2** and note when it was over $80. That step resulted a vector (even though you may not have thought of it as a vector) with two elements **[1, 4]**; that is, you noted that **priceStock_2** was over $80 at its first element and its fourth element. The second step you probably followed was to note that **priceStock_1** values at the first and fourth elements were **[25.11, 23.99]**. Even if you did not follow these two steps, convince yourself that this breakdown of the solution into these steps makes sense.

In MATLAB, you can use **find** to perform the first step (getting the indices where **priceStock_2** is greater than $80), and use the indices that **find** yields to access the right elements of **priceStock_1** in the second step:

```
>> ndxStock_2_Over80 = find(priceStock_2 > 80)
ndxStock_2_Over80 =
         1.00              4.00

>> valueStock1_whenStock2_Over80 =
priceStock_1(ndxStock_2_Over80)
valueStock1_whenStock2_Over80 =
         25.11             23.99
```

If you do not understand indexing into a vector using another vector, review Section 5-2.1.

You can use the same strategy to obtain the values of a vector that meet some test. Suppose you want to ask the following: "What are the stock prices of stock #1 when it is over $24. Apply the same two steps: Get the indices using **find** and a relational test; then, use the indices to index back into the price vector:

```
>> ndxStock_1_Over24 = find(priceStock_1 > 24)
ndxStock_1_Over24 =
         1.00              2.00              3.00

>> valueStock1_Over24 = priceStock_1(ndxStock_1_Over24)
valueStock1_Over24 =
         25.11             24.27             24.14
```

Use **find** in this way to give you the indices in a vector such that the values at those indices meets some relational test. Using the indices to dig out the values is a compact and useful operation.

The two examples above are examples of ***content addressing*** in computer science terms. Content addressing means accessing data by referring to its content and not its location. Here is another example of content addressing that will help you see its usefulness.

Suppose we keep records of company employees in the following vectors:

```
>> empSocSecNum = [111558298, 999551111, 787321479];
>> empBldgLocation = [1, 22, 1];
>> empPayRate = [32, 23, 55];
```

The **empSocSecNum** is the ID number for each employee and uniquely identifies the person. The **empBldgLocation** is the building a given individual works in. And **empPayRate** is the hourly rate of pay for each employee. So, what kind of questions can we ask? Below are some examples, along with a few lines of code to answer each.

- In which building does the employee with social security number 111558298 work?

```
>> ndxEmployee = find(empSocSecNum == 111558298);
>> bldgEmployee = empBldgLocation(ndxEmployee)
bldgEmployee =
          1.00
```

- Which employees work in Building 1?

```
>> ndxBldg = find(empBldgLocation == 1);
>> employeeInBldg = empSocSecNum(ndxBldg)
employeeInBldg =
  111558298.00  787321479.00
```

- Which employees make more than $30 per hour?

```
>> ndxEmployeesRateOver = find(empPayRate > 30);
>> employeesRateOver = empSocSecNum(ndxEmployeesRateOver)
employeesRateOver =
  111558298.00  787321479.00
```

Each of the examples above follows the same pattern for content addressing: Find the indices of the elements to be accessed, and use the indices to perform the access.

You should contrast content addressing with standard addressing. In MATLAB, you have learned to access vectors by specifying an index. An index can be a number, a vector of numbers, or some expression that when evaluated yields a number or a vector of numbers. You can call this *standard addressing*. Utilizing standard addressing, you tell MATLAB the element(s) of a vector you want to access by specifying a position in MATLAB memory. But you have learned to use the **find** built-in function to give you an index or indices in a vector such that specified tests are true for the contents of the elements at that index or those indices.

Using the index that **find** hands you, it is then possible to access other associated vectors. In the above example, you could **find** the building an employee works in given the employee's social security number. If you think of the vectors storing social security numbers and buildings as one database of information about employees, then you see that the access operations using **find** allow you to access parts of the database given contents of other parts of the database. This can be thought of as content addressing of data. MATLAB is one of the few programming environments that supplies the user an operation like **find** and supports content addressing in a direct fashion.

5-4.3   Vector Expressions and Rules for Forming Vector Expressions

In Section 3-4.2 you learned about constructing scalar expressions in MATLAB and the rules for making such expressions. All that you learned in Section 3-4.2 still holds for vectors. All we need to do is update the table showing precedence levels: This update is in Table 5-7. Items new in the precedence table to extend it to treat scalars and vectors are the following:

- the transpose operator (at level 2)
- the colon operator for vector creation (at level 6)
- cell-by-cell operations throughout the table

In Section 2-2, we gave a solution to the "hay bale problem." The crux in the hay bale problem was to determine the firing angle that would maximize the horizontal travel distance of a projectile.

The solution we developed involved algebraically manipulating background physics knowledge to develop an expression for the horizontal distance a projectile travels versus the initial firing angle of the projectile and the initial projectile speed:

$$D = \frac{2v_0^2 \sin(\phi)\cos(\phi)}{g} \qquad \text{(EQ 5-2)}$$

Table 5-7: Precedence Rules for MATLAB Operators (scalars and vectors, cell-by-cell only)

Precedence Level	Operator
1	parentheses
2	transpose (**'**) cell-by-cell exponentiation (**. ^**)
3	cell-by-cell unary plus (**+**) cell-by-cell unary minus (**−**) logical negation (**~**)
4	cell-by-cell multiplication (**. ***) cell-by-cell right division (**. /**)
5	cell-by-cell addition (**+**) cell-by-cell subtraction (**−**)
6	the colon operator (**:**)
7	cell-by-cell relational *all the relational operators* (**<, <=, >, >=, ==, ~=**)
8	cell-by-cell logical AND (**&**)
9	cell-by-cell logical OR (**|**)

v_0 is the initial projectile speed, g is the acceleration due to gravity, and ϕ is the firing angle. With Equation 5-2, we took a sample set of values of ϕ [0, 10, 20, …, 90 degrees] and constructed a two-column table with our values of ϕ and the corresponding values of D given a value of v_0. For an initial firing speed of 50 ft/sec, the data for the horizontal travel distance of a projectile for varying firing angles are shown in Table 5-8.

From the data in Table 5-8 (or its generalization in Table 2-2) what is our next step to get the angle that produces the maximum horizontal flight distance? The answer is to examine the table: Find the value in Column 1 data that maximizes Column 2 data. Or, plot the data in the table and find the maximum. For now, let's play out the first solution possibility utilizing your knowledge of vector operations in MATLAB to obtain the numbers for D given ϕ.

First, define a vector of values for ϕ. Then, by using cell-by-cell vector operations, compute a vector of corresponding D values. To be consistent with the data in Table 5-8, we will take the initial firing speed to be 50 ft/sec:

```
>> phi = linspace(0,90,10)';   % degrees
```

Table 5-8: Horizontal Distance ($v_0 = 50$ ft/sec)

ϕ (degrees)	D (feet)
0	0.0
10	26.7
20	50.2
30	67.6
40	76.9
50	76.9
60	67.6
70	50.2
80	26.7
90	0.0

The transpose operation is to convert the result from **linspace** to a column vector so that those results look more like Column 1 in Table 5-8.

Constructing an independent vector variable with **linspace** in MATLAB is an operation that you will use again. Since trig functions in MATLAB expect angle arguments to be in radians, convert ϕ to radians and translate Equation 5-2 into a MATLAB cell-by-cell vector computation:

```
>> initialFiringSpeed = 50;        % feet per sec
>> g = 32;                         % feet per sec per sec - gravity
>> phiRad = (pi / 180) * phi;      % radians
>> D = (2*initialFiringSpeed^2/g) .* sin(phiRad) .* cos(phiRad)   % feet
D =

            0
        26.72
        50.22
        67.66
        76.94
        76.94
        67.66
        50.22
        26.72
         0.00
```

The numbers for D in Table 5-8 were obtained with the code above although the numbers in Table 5-8 are shown to more significant digits.

Suppose you want greater precision in your results for D. How can you change the code above so you will get a better answer? The goal is to find the value of ϕ that produces the maximal value of D. By looking at the values of D computed, you can argue that the maximal value of D is between **D(5)** and **D(6)**, and hence the value of ϕ we are looking for is between **phi(5)** and **phi(6)**:

```
>> D(5), D(6)
ans =
          76.94
ans =
          76.94
>> phi(5), phi(6)
ans =
          40.00
ans =
          50.00
```

Because the values **D(1:5)** are a mirror image of the values **D(6:10)**, it's reasonable to assume that the value of **x** we seek is halfway between **phi(5)** and **phi(6)**, or 45 degrees.

But we are assuming a well-behaved function for **D**. How could we better test the assumption, i.e., how can we do the calculations with more precision? The answer is use more data points in the independent variable **phi**. By changing one line of the code above, we can accomplish that task:

```
>> phi = linspace(0,90)';  % degrees
```

If you do not specify the third argument to **linspace**, MATLAB will assume you want 100 equally spaced intervals in the vector you create. If you make this change to **phi**, recompute **D** and ask for **D(45:55)**. You will find the following answer:

```
>> D(45:55)
ans =
          76.94
          77.33
          77.64
          77.88
          78.04
          78.12
          78.12
          78.04
          77.88
          77.64
          77.33
```

The maximal value of **D** is now between **D(50)** and **D(51)**. Here are the corresponding values of **phi**:

```
>> phi(50:51)
ans =
        44.55
        45.45
```

The answer that maximizes **D** is 45 degrees. If you want more accuracy to further convince yourself that 45 degrees is the right answer, you could use 1000 as the third argument to **linspace**, in which case you would find the maximal value of **D** to be between **D(500)** and **D(501)**, with the values of **phi** producing the two following values of **D**:

```
>> phi(500:501)
ans =
        44.95
        45.05
```

For the projectile problem, the solution path was to find a numerical solution.[7] You had a relationship between an independent variable (ϕ) and a dependent variable (**D**). You wanted the value of ϕ that would maximize **D**. Instead of inverting the relationship (Equation 5-2) to get an analytical solution, you created a set of values of ϕ and, for each, found the corresponding values of **D**. After, you examined the set of **D** values to find the maximal value, and picked off the value of ϕ, corresponding to the maximal value of **D**. You didn't exercise everything you have learned in this section about vectors, but you have exercised some essential capabilities.

5-4.4   Section Synopsis

The main point of this section has been to extend your knowledge of MATLAB to vectors. Like all facets of a computational environment, you cannot effectively learn how to apply vector cell-by-cell operations by reading about it. You must practice all the problems at the end of this section. Below is a synopsis of material in this section.

---

7. In this problem, you could have found an analytic solution by differentiating Equation 5-2 with respect to ϕ and setting the result to 0. But many times in engineering, an analytical solution to a problem does not exist.

1. Two major classes of built-in functions in MATLAB operate on vectors: functions that work in a cell-by-cell fashion (like sin) and functions that aggregate (like sum).

2. Cell-by-cell vector operators take two arguments and apply the indicated operation (addition, subtraction, etc.) to the corresponding elements of the two vectors to produce the result vector.

3. For cell-by-cell operations, the two arguments must be the same type of vector (row or column) and be of the same length, or one of the arguments must be a scalar.

4. If the application of an operator has one argument being a scalar and another being a vector, an easy way to understand the result is to convert the scalar to a vector of like kind and length to the other argument, and to perform a standard cell-by-cell operation.

5. Cell-by-cell vector operators include the classes (assignment, colon and transpose operators), the vector arithmetic cell-by-cell operators (Table 5-2), the vector relational operators (Table 5-4), and vector cell-by-cell logic operators (Table 5-5).

6. Logical computations are extended *via* built-in logic functions (Table 5-6).

7. The built-in logic function find is useful because it enables a type of content addressing.

8. The operator precedence table was updated to include new possibilities (Table 5-7).

9. A common operation in MATLAB is to designate an independent vector variable, create representative values for the elements of that independent variable, and compute values for a dependent vector variable using a known relationship.

Synopsis for Section 5-4

5-5   Problem Sets for Vectors

5-5.1   Set A: Nuts and Bolts Problems for Vectors

Problem 5-A.1 (Section 5-1)

Create a row vector **a1** consisting of the numbers in the ordered set {1, 5, 7, 22, 12} using the square bracket operator.

Problem 5-A.2 (Section 5-1)

Create a row vector **a2** consisting of the numbers in the ordered set {1, 2, 3, 4, 5} using the square bracket operator.

Problem 5-A.3 (Section 5-1)

Create a row vector **a3** consisting of the numbers in the ordered set {1, 2, 3, 4, 5} using the colon operator.

Problem 5-A.4 (Section 5-1)

Create a row vector **a4** consisting of the numbers in the ordered set {1, 4, 7, 10} using the square bracket operator.

Problem 5-A.5 (Section 5-1)

Create a row vector **a5** consisting of the numbers in the ordered set {1, 4, 7, 10} using the colon operator.

Problem 5-A.6 (Section 5-1)

Create a row vector **a6** starting with the element 1, ending with the element 10, with five equally spaced points in the vector. Use `linspace`.

Problem 5-A.7 (Section 5-1)

Create a row vector **a7** starting with the element 1, ending with the element 10, with five equally spaced points in the vector. Use the colon operator.

Problem 5-A.8 (Section 5-1)

Create a row vector **a8** starting with the element 1, ending with the element 10, with five equally spaced points in the vector. Use the square bracket operator.

Problem 5-A.9 (Section 5-1)

Build a row vector **a9** that has ten elements, each with the value 1. Use the MATLAB built-in function **ones**.

Problem 5-A.10 (Section 5-1)

Build a row vector **a10** that has five elements, each with the value 33. Use the built-in function **ones**.

Problem 5-A.11 (Section 5-2)

In MATLAB, define variable **x** as follows:

```
>> x = [3, 9, 4, 99, 2]
```

Set a variable **y** to be the number of elements in **x** – use **length**.

Problem 5-A.12 (Section 5-2)

In MATLAB, define variable **x** as follows:

```
>> x = [3, 9, 4, 99, 2]
```

Set a variable **y** to be the first and fourth elements of **x**.

Problem 5-A.13 (Section 5-2)

In MATLAB, define variable **x** as follows:

```
>> x = [3, 9, 4, 99, 2]
```

Set a variable **y** to be the first, second, and third elements of **x**. Use the colon operator in your solution.

Problem 5-A.14 (Section 5-2)

In MATLAB, define variable **x** as follows:

```
>> x = [3, 9, 4, 99, 2]
```

Set a variable **y** to be the third through the last element of **x** so your solution works no matter how many elements are in **x**.

Problem 5-A.15 (Section 5-2)

In MATLAB, define variable **x** as follows:

```
>> x = [3, 9, 4, 99, 2]
```

Set a variable **y** to be the next-to-last and the last element of **x** so your solution works no matter how many elements are in **x**.

Problem 5-A.16 (Section 5-2)

In MATLAB, define variable **x** as follows:

```
>> x = [3, 9, 4, 99, 2]
```

Change the second element of **x** to be 3.

Problem 5-A.17 (Section 5-2)

In MATLAB, define variable **x** as follows:

```
>> x = [3, 9, 4, 99, 2]
```

Change the second element of **x** to be 102 and the fourth element of **x** be 205.

Problem 5-A.18 (Section 5-2)

In MATLAB, define variable **x** as follows:

```
>> x = [3, 9, 4, 99, 2]
```

Change the last element of **x** to be 999 so that your solution works no matter the length of **x**.

Problem 5-A.19 (Section 5-2)

In MATLAB, define variable **x** as follows:

```
>> x = [3, 9, 4, 99, 2]
```

Change the next-to-last element of **x** to be 55 and the last element to be 65 so that your solution works no matter the length of **x**.

Problem 5-A.20 (Section 5-2)

In MATLAB, define variable **x** as follows:

```
>> x = [3, 9, 4, 99, 2]
```

Delete the next to last element of **x** so it will leave **x** as a vector with only four elements.

Problem 5-A.21 (Section 5-3)

In MATLAB, define variable **x** as follows:

```
>> x = [3, 9, 4, 99, 2]
```

Change **x** into a column vector.

Problem 5-A.22 (Section 5-3)

Create a column vector **y** whose elements are {2, 40, 1} using the square bracket operator.

Problem 5-A.23 (Section 5-3)

Using the **x** vector result from Problem 5-A.21 and the **y** vector result from Problem 5-A.22, create a column vector **z** whose first five elements are from **x** and whose next three elements are from **y**.

Problem 5-A.24 (Section 5-3)

Change the **z** vector result from Problem 5-A.23 to a row vector.

Problem 5-A.25 (Section 5-4)

Define **x** as follows:

```
>> x = [3; 9; 4; 5; 2]
```

Assume that values in **x** are angles in radians. Create a vector **y** such that each element in **y** is the sine of the corresponding element in **x**.

Problem 5-A.26 (Section 5-4)

Define **x** as follows:

```
>> x = [3; 9; 4; 5; 2]
```

Create a vector **y** so each element y_i has the value e^{x_i} for the corresponding i^{th} element in **x**.

Problem 5-A.27 (Section 5-4)

Define **x** as follows:

```
>> x = [3; 9; 4; 5; 2]
```

Create a variable **s** which holds the sum of the elements of **x**.

Problem 5-A.28 (Section 5-4)

Define **x** as follows:

```
>> x = [3; 9; 4; 5; 2]
```

Create a variable **myMax**, which is the maximum value of all elements in **x**.

Problem 5-A.29 (Section 5-4)

Define **x** as follows:

```
>> x = [3; 9; 4; 5; 2]
```

Create a variable **myMean** which is the arithmetic mean of all scalar elements in **x**.

Problem 5-A.30 (Section 5-4)

Define **x** as follows:

```
>> x = [3; 9; 4; 5; 2]
```

Create a variable **mySort**, which contains all values in **x** in sorted order.

Problem 5-A.31 (Section 5-4)

In MATLAB, define variables **x** and **y** as follows:

```
>> x = [3; 9; 4; 5; 2]
>> y = [2; 1; 4; 3; 2]
```

Create a new vector **z** in which each element is the sum of the corresponding elements in **x** and **y**.

Problem 5-A.32 (Section 5-4)

In MATLAB, define variables **x** and **y** as follows:

```
>> x = [3; 9; 4; 5; 2]
```

```
>> y = [2; 1; 4; 3; 2]
```

Create a new vector **z** in which each element is twice the value of the corresponding element in **x**, added to three times the corresponding value in **y**.

Problem 5-A.33 (Section 5-4)

In MATLAB, define variables **x** and **y** as follows:

```
>> x = [3; 9; 4; 5; 2]
```

```
>> y = [2; 1; 4; 3; 2]
```

Create a new vector **z** in which each element is the product of the corresponding elements in **x** and **y**.

Problem 5-A.34 (Section 5-4)

In MATLAB, define variables **x** and **y** as follows:

```
>> x = [3; 9; 4; 5; 2]
```

```
>> y = [2; 1; 4; 3; 2]
```

Create a new vector **z** in which each ith element in **z** is $(x_i)^{y_i}$.

Problem 5-A.35 (Section 5-4)

In MATLAB, define variables **x** and **y** as follows:

```
>> x = [3; 9; 4; 5; 2]
```

```
>> y = [2; 1; 4; 3; 2]
```

Create a new vector **z** in which each element is three times the value of the corresponding value in **x**, divided by five times the value of the corresponding value in **y**.

Problem 5-A.36 (Section 5-4)

In MATLAB, define variables **x** and **y** as follows:

```
>> x = [3; 9; 5; 5; 1]
```

```
>> y = [2; 1; 4; 6; 2]
```

Are any elements in **x** such that corresponding elements of **y** are twice as big?

Problem 5-A.37 (Section 5-4)

In MATLAB, define variables **x** and **y** as follows:

```
>> x = [3; 9; 5; 5; 1]
>> y = [2; 1; 4; 6; 2]
```

Are all elements in **x** such that corresponding elements of **y** are twice as big?

Problem 5-A.38 (Section 5-4)

In MATLAB, define variables **x** and **y** as follows:

```
>> x = [3; 9; 5; 5; 1]
>> y = [2; 1; 4; 6; 2]
```

Find the indices for all elements in **x** such that corresponding elements of **y** are twice as big.

Problem 5-A.39 (Section 5-4)

In MATLAB, define variables **x** and **y** as follows:

```
>> x = [3; 9; 5; 5; 1]
>> y = [2; 1; 4; 6; 2]
```

Find the indices for elements in **x** so the corresponding elements of **y** are twice as big as the element in **x** or the corresponding elements of **y** are less than **x**.

Problem 5-A.40 (Section 5-4)

In MATLAB, define variables **x** and **y** as follows:

```
>> x = [3; 9; 5; 5; 1]
>> y = [2; 1; 4; 6; 2]
```

Find the indices for all elements in **x** so neither the corresponding elements of **y** are twice as big as the element in **x** nor the corresponding elements of **y** are less than **x**.

Problem 5-A.41 (Section 5-4)

In MATLAB, define variables **x** and **y** as follows:

```
>> x = [3; 9; 5; 5; 1]
```

```
>> y = [2; 1; 4; 6; 2]
```

Find the locations and values of all the elements in **x** that have values between 4 and 10.

Problem 5-A.42 (Section 5-4)

In MATLAB, define variables **x** and **y** as follows:

```
>> x = [3; 9; 5; 5; 1]
```

```
>> y = [2; 1; 4; 6; 2]
```

Set all the elements in **x** that have values between 4 and 10 to the value of the corresponding value in **y**.

Problem 5-A.43 (Section 5-4)

In MATLAB, define variables **x** and **y** as follows:

```
>> x = [3; 9; 5; 5; 1]
```

```
>> y = [2; 1; 4; 6; 2]
```

Set all the elements in **y** that have values greater than or equal to 3 to two times their initial value.

Problem 5-A.44 (Section 5-4)

In MATLAB, define variables **x** and **y** as follows:

```
>> x = [3; 9; 5; 5; 1]
```

```
>> y = [2; 1; 4; 6; 2]
```

For all the elements in **x** such that *either* the corresponding elements of **y** are twice as big as the element in **x** *or* the corresponding elements of **y** are less than **x**, set those elements in **x** to the number 99.

Problem 5-A.45 (Section 5-4)

The relationship $dist = \frac{1}{2}at^2$ gives the distance traveled ($dist$) by a particle

starting from standstill, undergoing uniform acceleration (a) after some time (t). Create a vector t having 100 equally spaced values ranging from 0 to 50 seconds. Assume that a is 100 m/s/s. Compute a vector **dist** with the same number of elements as **t**, and for which each element is the distance traveled for the corresponding time of travel in **t**.

5-5.2   Set B: Problem Solving with MATLAB Vectors

Problem 5-B.1

The melting temperature is a good indicator to discover the identity of an unknown substance. Suppose we perform four independent experiments and find melting temperatures of four samples to be [848 854 851 852] °C. A second round of experiments gives the following results: [850 853 849 851] °C.

Store the results of the two rounds of experiments in vectors **firstRoundResults** and **secondRoundResults**, respectively.

Now define a new vector **allResults** by concatenating **firstRoundResults** and **secondRoundResults**. Check the lengths of all three vectors and verify the length of **allResults** is the sum of the lengths of **firstRoundResults** and **secondRoundResults**.

Problem 5-B.2

Part A   Create a vector **x** using the square bracket method that has the following elements:

{0 3 6 9 … 81 84 82 80 78 76 … 0}

Part B   Create a vector **x1** using the colon operator that has the following elements:

{0 3 6 9 … 81 84}

Part C   Create a vector **x2** using the colon operator that has the following elements:

{82 80 78 76 … 0}

Part D   Create a vector **x_AnotherWay** that is identical to **x** by concatenating your result from **Parts B** and **C**.

Problem 5-B.3

What does MATLAB return when you execute the following line in the MATLAB Command Window?

```
>> x = [.7:5 linspace(1,6,6)]
```

Figure this out with paper and pencil first, then run the command in MATLAB to verify your result.

Problem 5-B.4

Use two methods to create a vector with 30 elements, where each element is the number 30. Use built-in functions **zeros** in one method and **ones** in the other method.

Problem 5-B.5

Create the following vectors using the colon operator and the built-in function **linspace** methods:

Part A   A vector whose elements are all odd numbers between 1 and 191

Part B   A vector whose elements are 99 evenly spaced numbers between 5 and 148

Problem 5-B.6

We would like to display a temperature given in $^{\circ}$F on three temperature scales, $^{\circ}$F, $^{\circ}$C, and $^{\circ}$K, in a vector of length 3. The conversion from $^{\circ}$F to $^{\circ}$C and from $^{\circ}$C to $^{\circ}$K are given as follows:

$$tempCelsius = \frac{tempFahrenheit - 32}{1.8}$$

$$tempKelvin = tempCelsius + 273.1$$

Write a MATLAB function that accepts a scalar for temperature in Fahrenheit degrees as input and outputs a vector of degrees in $^{\circ}$F, $^{\circ}$C, and $^{\circ}$K.

Problem 5-B.7

Two vectors, **x** and **y**, are given as follows:

```
x = [8 4 6 5 9 3]

y = [1 2 3 4 5 6]
```

Fetch every other element in **x** starting with the second element. Use the fetched elements to replace the last three elements of **y**.

Problem 5-B.8

Write a MATLAB function that takes a positive, non-zero integer **N** and a vector **V** as inputs and produces a single output vector such that every N^{th} element in **V** is set to the value –99.

For example, if **N=3**, and **V=[1 2 3 4 5 6 7]**, the output returned would be **[1 2 -99 4 5 -99 7]**.

Problem 5-B.9

Find the values of **z** in the following questions with paper and pencil first. When finished, execute the commands to verify your answers.

Part A   `x = linspace(2,20,10);z = x([3:3:length(x)])`

Part B   `x = [7 3 4 2 35 23];z = [x([4 5]), x([2 1])]`

Part C   `z = [1:10];z([6:length(z)]) = z([1:5])`

Part D   `z = [4 8 2 1];z([1, length(z)]) = [1 4]`

Problem 5-B.10

Use vector **x** = [4 8 2 1]. Write a MATLAB function that takes no inputs and outputs a shuffled version of **x**. The function will shuffle the elements in vector **x** by performing the following steps:

     1. Switch the first and last elements of **x**.

     2. Switch the first two elements with the last two elements in **x**.

For the value for **x** above, the output of your function should be [2 4 1 8].

Problem 5-B.11

Write a MATLAB function that accepts two integers, a and b, and outputs a column vector that is composed of every other number between a and b.

Problem 5-B.12

Write a MATLAB function that accepts two scalars, a and b, and an integer c, and outputs a column vector that is composed of c evenly spaced numbers between a and b.

Problem 5-B.13

Create the following vectors using first **linspace** or the colon operator, then the transpose operator.

Part A   A column vector of all even numbers between (and including) 0 and 10

Part B   A column vector of five evenly spaced numbers between (and including) 2 and 10

Part C $\begin{bmatrix} 4 \\ 7 \\ 10 \\ 13 \\ 16 \end{bmatrix}$

Problem 5-B.14

A two-dimensional physics vector can be represented in MATLAB by a collection of its x and y components. For example, V_1 and V_2 in the figure below could be represented as (4,1) and (3,4), respectively. The summation of two vectors is carried out by adding the components of vectors independently. In the example below, V_R, which is the sum of V_1 and V_2, is calculated as (4+3, 1+4). If vectors represent forces, a summation of vectors is said to be the resultant force.

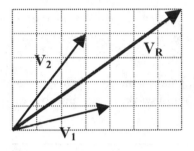

Write a MATLAB function that accepts two arbitrary forces in a two-dimensional space and returns the resultant force.

Problem 5-B.15

When an object of mass m attached to a spring from above is free, two relations for the two forces are acting on the spring: the gravitational force giving rise to the weight of the mass (pulling the mass downward) and the force exerted by the spring (pulling the mass upward). At rest, the two forces must be equal to each other in magnitude but opposite in sign:

$$F_{gravitational} = -mg$$
$$F_{spring} = -k_{spring}x$$
$$\therefore mg = -k_{spring}x$$

The first relation gives the force due to gravity (the weight). The second relation gives the restoring force due to the spring, where k_{spring} is the spring constant, i.e., a constant of proportionality that is descriptive for one given spring and where x is the length that the weight stretches the spring. The third relation puts it all together since the spring force and the gravitational force are equal in magnitude but opposite in direction.

We will take the units for the relation $mg = -k_{spring}\,x$ to be the following: unit for mass (m) is kilograms, unit for displacement (x) is meters, unit for the spring constant (k) is Newtons/meter. The value of g, the acceleration due to gravity, is 9.81 m/s².

The potential energy stored in the string due to this elongation is known to be (in Newton-meters):

$$potentialEnergy = \frac{kx^2}{2}$$

Write a MATLAB function that accepts a vector of extension distances (**x**) and a vector of corresponding spring constants (**k**). Your function should return a vector of object masses and a vector of potential energies stored in each spring.

Test your function with data for the five spring-mass examples in the figure below. Your function should work for any number of spring-mass systems.

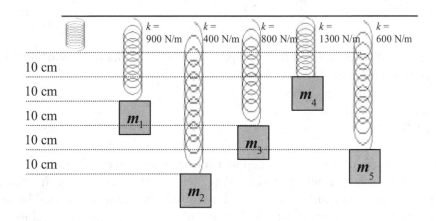

Problem 5-B.16

Suppose you have a catalogue of automobiles such as in the following table:

Write a MATLAB function that will input a vector for **ID**, a vector for **Color**, a vector for **Price**, and return the ID numbers of black cars that are priced under $30,000.

Problem 5-B.17

When a light source is submerged in a pond and pointed toward the surface of the pond, either reflection or refraction occurs when the light beam crosses from under the water to above the water. Whether refraction or reflection takes place

ID	Color (0 - black, 1 - blue, 2 - red)	Price($)
1002	0	24,000
2176	1	18,000
3201	2	36,000
4204	1	26,000
5512	2	35,000
6309	0	29,000
8841	1	31,000
9004	0	38,000

depends on the incidence angle α_1 of the light beam. The physical situation is shown in the following diagram:

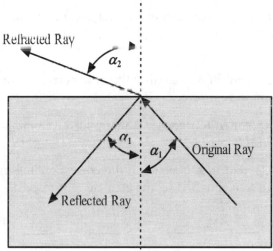

When α_1 is smaller than *the critical angle*, the light beam will undergo *both* refraction and reflection. That is, some of the light will be passed out of the water and into the air, but its direction will change (refraction), and some of the light will be bounced from the surface back into the water (reflection).

Refraction is described according to Snell's law in the following relation:

$$n_1 \sin(\alpha_1) = n_2 \sin(\alpha_2)$$

where n_1 is the index of refraction of water (1.33), and n_2 is the index of refraction of air (1.00029), and the angles are shown in the diagram above.

If angle α_1 is larger than the critical angle, only internal reflection occurs, and the angle of reflection will be the same as the incidence angle as shown in the diagram above.

The critical angle for the water-air boundary is 49°. Write a MATLAB function that accepts a vector of incidence angles (α_1, in degrees) and returns a vector of reflection or refraction angles (in degrees) for each incidence angle. Assume that for angles smaller than the critical angle only refraction occurs. Try your function with incidence angles from 0 to 90° with increments of 5°.

Given the vector of angles, find which angles are greater than critical angle first. Calculate refraction angles and replace some refraction angles with reflection angles.

Problem 5-B.18

Write a function that will take vectors **A** and **B** as input, compare and update the two vectors according to the following rules, and output the updated vectors **A** and **B**.

1) If an element in **A** is smaller than the corresponding element in **B**, set the element in **A** to zero.

2) If an element in **A** is greater than the corresponding element in **B**, set the element in **B** to zero.

3) If an element in **A** is equal to the corresponding element in **B**, set the element in **A** to zero and set the element in **B** to zero.

Problem 5-B.19

A farmer has a square shaped field with side length L ft. He then buys another field which is square, but its side length is $L+200$ ft. He knows that the total area of the two fields is between 110 acres and 120 acres. (1 sq. mi = 640 acres; 5280 ft = 1 mi)

Write a MATLAB function that accepts a two-element vector for the range of total area (in acres) as input (`[110, 120]` in this case), computes a vector of possible values of L (in feet), and returns a vector of possible values of L. Consider integer values of L between 100 ft and 5,000 ft in increments of 100 ft.

Chapter 6

2-D and 3-D Plotting and Using MATLAB Help

In Chapter 5, you extended your MATLAB capabilities to the features that make MAT-LAB distinctive among computer environments for technical problem solving. Vectors in MATLAB are ordered groups of scalar numbers.[1] Vector operations enable you to perform two types of computations. First, you can apply the same operation to all elements in a vector. An example of this would be applying a built-in function like **sin** to a vector (NOTE: Be sure to read footnote (2) at the bottom of this page!):

```
x = 0:pi/4:2*pi;[2]
sinX = sin(x);
```

Second, vector operations enable you to accumulate an answer based on the elements in a vector. An example would be the application of the built-in function **mean** to a vector to obtain the scalar value for the arithmetic average of the elements of the vector:

```
y = [23.2, 35.1, 10.2];
averageOfY = mean(y);
```

Another example would be applying the built-in function **sort** to a vector to produce another vector that has all elements of the first, but in sorted order:

```
y = [23.2, 35.1, 10.2];
sortedY = sort(y);
```

---

1. MATLAB supports vectors which are ordered groups of other objects, typically strings but with the possibility of being any other object type supported by MATLAB. These extended vectors, extended beyond holding scalars, are called *cell vectors*. If you want to know more about cell vectors (or cell arrays), use MATLAB Help.

2. From this point on, we will usually avoid using the **>>** when we display code. Those symbols have indicated MATLAB commands typed into the Command Window. All code lines from this point are printed without **>>** to indicate the command is valid whether appearing as a single line typed in the Command Window or as a line in a function file. The exception will be when we want to indicate a command typed to the Command Window followed by the result that MATLAB produces. In this case, we will use the **>>**.

Other computational environments, such as MathCad or Mathematica, support vector operations, but MATLAB is the oldest and best-developed system that enables such compact and versatile operations on an ordered set of scalars.

We have one more major piece to develop in your understanding of the fundamentals of MATLAB: Matrix and Array Operations. Before moving to arrays in Chapter 7, in this chapter we will take a little break from introducing new core concepts and discuss two topics: 2-D and 3-D Plotting with MATLAB, and the use of the **Help** facilities in MATLAB.

MATLAB offers a full set of functions for creating plots. In addition to the wide range of plot types available, one of the best aspects about creating plots in MATLAB is that it is straightforward once you have mastered the concept and use of vectors.

The second major topic of this chapter is the use of the Help Facility in MATLAB. Gaining proficiency in the use of MATLAB's Help is crucial. MATLAB has more capability than can possibly be covered in a single introductory course. But happily, you don't have to know every single part of MATLAB to be an efficient user of MATLAB. After you master the basics, you may find that you need help with some MATLAB function or capability you have or have not used before. MATLAB Help will, in fact, help you in such situations.

Rather than introduce MATLAB Help as a separate topic, in the remainder of this chapter, Help will be demonstrated as subjects on plotting are introduced.

6-1   Using ezplot to Graph Functions

MATLAB includes a built-in function called **ezplot** (pronounced "easy plot"). You use **ezplot** much as you would a graphing calculator, like the TI-83. As the function's name implies, **ezplot** is the quick and dirty plotting function of MATLAB.[3] In this section, you will learn to use **ezplot** to (1) plot a function of one variable, (2) plot implicitly defined functions of two variables, and (3) plot two functions each defined in terms of the same variable. In more mathematical terms, the three cases are as follows.

1.   Single function of one variable: $f = f(x)$
2.   Single function of two variables that is equal to zero: $g = g(x, y) = 0$
3.   Two functions both defined over a single independent variable: $h = h(t)$, $k = k(t)$

6-1.1   Using ezplot for a Function of One Variable and Getting Help

Before we turn full attention to **ezplot**, this is a good time for you to begin learning to use MATLAB Help. As pointed out earlier, you can never hope to learn everything there is to know about MATLAB from an introductory book. You need to develop the ability to augment what you already know about MATLAB. One of the easiest ways to learn about new features of MATLAB is by referring to MATLAB's Help.

To open Help, click on the question mark icon in the Command Window as shown in Figure 6-1.[4] MATLAB Help offers a number of ways of finding information about MATLAB use. You will learn about some of the various methods in this chapter.

If you know the name of a function you need help with you can find information most quickly by letting MATLAB know what you are looking for. One example is when you want to find information about a function whose name you already know. When you first open the Help window, you will see two panes as shown in Figure 6-2. One of the panes is used for navigation, i.e., for finding what you want to find in Help. The

---

3.   **ezplot** is part of an extensive capability in MATLAB that deals with symbolic mathematics. The "Symbolic Toolbox" will be the topic of Chapter 7.

4.   There are other ways of bringing up the Help facility. For example, in the Command or Edit window, if you highlight a MATLAB built-in function name and right click, you will get a menu that includes the selection Help ON SELECTION. This option brings up Help with the function you highlighted shown and is a quick way of getting Help for functions for which you know the name.

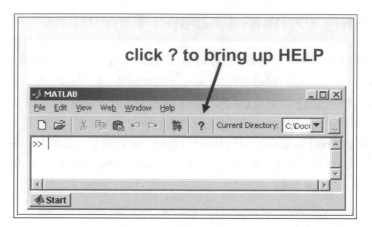

Figure 6-1: Initiating MATLAB Help from the Command Window

other pane displays information about the topic you select in the navigation pane.

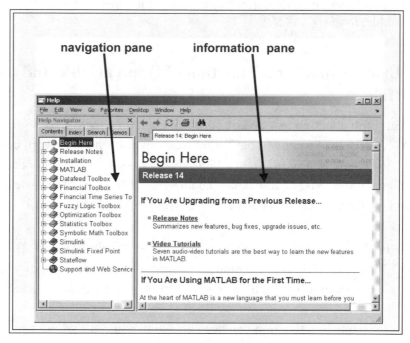

**Figure 6-2: Help Window Panes Showing Navigation
and Information**

Click the tab in the navigation pane labeled *Search*. Then type into the Search field the name **ezplot**. Hit Return on your keyboard, or click on Go to the right of the Search field. You will see the MATLAB Help entry for **ezplot** come up in the information pane.

Keep the Help window with the **ezplot** information up. Notice that the information about **ezplot** is shown in groups. The first group is simply a sentence about what

ezplot's purpose is. The second group tells you how to write code using **ezplot** (Syntax). Multiple forms are shown for **ezplot**. The third group is a detailed explanation of each syntactic form (Description). The fourth group lists common pitfalls for using **ezplot** (Remarks), and the last grouping lists some examples for the use of **ezplot** (Examples).

The Help description of **ezplot** tells you that **ezplot** is used for function plotting. The Syntax section tells you the forms you can use when calling **ezplot**. The first two forms shown in the Help Syntax for **ezplot** apply for a function of one variable.

The first form assumes a default upper and lower value for the domain of the function; the default limits are $[-2\pi, 2\pi]$. The syntax for this form is easy. If you want to plot the sine of a variable x and you will accept the defaults on the upper and lower limits of x, then the call to **ezplot** is the following:

```
ezplot('sin(x)')
```

The one input argument to this form is the function we want to have plotted, set off in single quotation marks.[5] We could have achieved the same thing by doing the following:

```
xx = 'sin(x)';
ezplot(xx)
```

Here is another example, one a little more interesting; execute the following in MATLAB:

```
yy = 'sin(3*x) / cos(x)';
ezplot(yy)
```

Remember, this first form assumes that the independent variable for your target function is defined in terms assumed to have a range of **x** that is $[-2\pi, 2\pi]$.

Suppose you want to set the range on your independent variable yourself. For example, for a function that represents the distance an object will travel in some time t under gravitational acceleration, $f(t) = \frac{1}{2}gt^2$, you would not choose to set values of t to range over negative values. In such cases, you would want to select the second form in the given Help/Syntax entry. Suppose you want to plot this function for values of t from 0 to 100 seconds. Following the second form, and knowing that g is 9.8 m/s², the call to ezplot is the following:

---

5. In Chapter 7 you will learn how to use ezplot without surrounding quotation marks – by using "symbolic variables."

```
>> ezplot('0.5 * 9.8 * t^2', [0, 100])
```

The numerical value of **g** had to be inserted as a scalar into the expression given to **ezplot**; you could not put it in as the variable **g** even if you had set **g** to 9.8 in the Command Window. In Chapter 7 you will see that there is a method to allow "parameterizing" an **ezplot** operation.

Several labeling functions can be used in MATLAB once a graph has been created, and they apply equally well to all MATLAB graphs, including those made with **ezplot**. Suppose we want to plot the function $f(x) = \dfrac{\sinh(3x)}{\cosh(x)}$. We want to label the x-axis, and y-axis, title the plot, and turn on a grid for the plot. All this is accomplished by the following:

```
figure                           % this creates a new plot
                                 % window
ezplot('sinh(3*x) / cosh(x)')    % puts the plot into the
                                 % new window
xlabel('angle (radians)')        % label the x-axis
ylabel('value of function')      % label the y-axis
title('sinh(3x)/cosh(x)')        % create a title
grid on                          % turn on a grid in the
                                 % plot
```

ezplot creates a label for the x-axis and a title for the graph. Often, we will want to augment the labeling to make a graph more understandable. Turning on the grid makes the graph results easier to read.

The labeling functions we used (**figure**, **xlabel**, **ylabel**, **title**, and **grid**) will be useful for all the plot types you encounter in MATLAB.

6-1.2   Using ezplot for a Function of Two Variables (Implicit form)

Refer back to the Help window with help for **ezplot** and focus on the Syntax section. The first three forms apply when we want to plot a function of two variables, like $f(x, y) = \sin(3x) - y^3$. Suppose we have x and y ranging over $[-2\pi, 2\pi]$. Using **ezplot** to bring up the plot is the easy part this time:

```
ezplot('sin(3*x) - y^3')
```

The plot window which MATLAB brings up will look like that shown in Figure 6-3.

For functions of two variables, such as $f(x, y) = \sin(3x) - y^3$, what is actually plotted is $f(x, y) = \sin(3x) - y^3 = 0$. The interpretation is: Find an (x,y) point on the curve

shown in the graph. The (x,y) pair of values satisfies the equation
$f(x, y) = \sin(3x) - y^3 = 0$. It is important to understand the difference between what is plotted for a function of a single variable versus what is plotted for a function of two variables such as that shown in Figure 6-3.

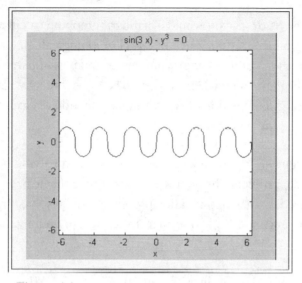

Figure 6-3: ezplot for a Function of Two Variables

Putting limits on the range of the two variables is straightforward (see Help/Syntax on **ezplot**). For example, in Figure 6-3, you can see that limiting the value of **y** to [–2, 2] and limiting **x** to a value of [0, 3] will produce a higher quality plot with no loss of information:

```
ezplot('sin(3*x) - y^3 ', [0,3,-2,2])
grid on
```

6-1.3   Using ezplot for Two Functions of the Same Variable (Parameterized form)

The fourth and fifth forms in Help/Syntax on **ezplot** are for cases in which you have two functions which are of the same independent variable: $f = f(t); \ g = g(t)$. Suppose we want to explore the relationship between a function $f = \sin(x)$ and another function $g = \cos(x)$. Suppose further we would like to know the value of g given a value for f. The fourth form of **ezplot** will give us the tools to answer this. In the Command Window do the following:

```
ezplot('cos(t)', 'sin(t)')
```

Using the plot that is generated, answer this question: What is the value of the sine when the value of the cosine is zero?

6-1.4   Synopsis

You can think of **ezplot** as you would a standard graphing calculator. The 2-D version of **ezplot** produces results much like a TI graphing calculator would, except in MATLAB you have the ability to copy your charts and paste them into documents such as reports. The main thing when using **ezplot** is deciding when you want to use the three different forms: function of one variable, function of two variables, or two functions both of the same variable.

Using **ezplot**, you can support a large range of problem solving based on charting. Back in the hay bale problem, the goal was to maximize the horizontal distance traveled by a projectile. Such optimization problems can be solved by creating a graph of the desired function and reading the chart to see where optimization points are reached.

Now that you are somewhat familiar with plotting in MATLAB using **ezplot**, the next step in learning about plotting in MATLAB will be to extend your knowledge to include plotting numerical data. The most important reason that this chapter is placed immediately after the chapter on vectors is that plotting numerical data in MATLAB is a piece of cake once vectors are mastered.

1. When you know the name of a function but not its details (like the arguments it takes for input), then use Help/Search/Function Name to find the details you seek.

2. Use `ezplot` for functions of one variable to make a quick and dirty functional chart. Optional arguments allow changing the default functional domain $[-2\pi, 2\pi]$.

3. Use `ezplot` for implicitly defined functions of two variables, e.g., $f(x,y) = 0$. Default ranges on x and y can be changed by using optional arguments.

4. Use `ezplot` for parametrically defined relationships $f = f(t)$, $g = g(t)$.

5. Use `xlabel`, `ylabel`, and `title` built-in functions to refine labeling the plots made by `ezplot`. Use the built-in function `grid` to activate a grid on a plot created. Use the built-in function `figure` to create a new plotting window.

Synopsis for Section 6-1

6-2   Using `plot` to Graph Numerical Data

Two distinct problem types exist for plotting numerical data using MATLAB. Both types use the same
basic plotting built-in functions, so what you need to
keep straight is the goals you have in creating a plot
and the necessary setup prior to using the built-in
functions for numerical plotting.

**Table 6-1: Speed and
Stopping Distance for
Braking System X2**

Speed (mi/hr)	Stopping Distance (ft)
20	46
30	75
40	128
50	201
60	292
70	385

The first type of problem in which you want to make
a numerical plot is when you have real numerical
data for the results of an experiment you have
conducted. Your goal would be to understand the
results of the experiment. The second type of
problem is when you want to visualize some
mathematical relationship, which is the same goal as
when you use **ezplot**.

The general goal to create a plot is the same
regardless of whether you are creating a numerical plot or a symbolic plot using
ezplot. The goal is to understand the relationship between variables in some
system. For example, if you are an automotive engineer conducting tests on a new
automotive braking system, you might have data such as that shown in Table 6-1. Your
goal is to understand the relationship between the independent variable in your
experiment (the speed) and the dependent variable (the stopping distance). The values
of the independent variable are under your control. In the example, you choose what
values of speed you will select for running a single experiment. The values of the
dependent variable are a result of the choices you make. In the example, the stopping
distance is dependent on the choice you made for the speed. Put another way, you
choose a set of test values for speed, and you measure the resultant set of stopping
distances.

Having done the experiment, and gotten the data in Table 6-1, your goal becomes to
understand the relationship between speed and stopping distance.

Your first step should be to make at least a rough plot of the data in a chart depicting
stopping distance versus speed like the one shown in Figure 6-4. Why?

Look at Table 6-1. From the data in the table, describe the relationship between
stopping distance and speed.

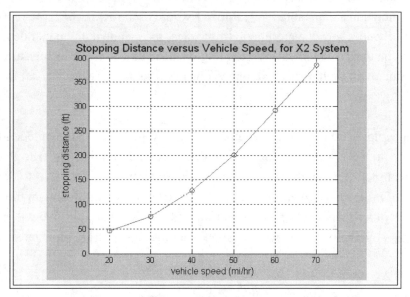

Figure 6-4: Plot for Data Shown in Table 6-1

Most likely your description was something like "The stopping distance goes up as velocity goes up." This is a correct description but incomplete. Look again at Figure 6-4, and describe the relationship between stopping distance and speed with more precision.

From the plot, you probably answered something like "The stopping distance goes up as the speed goes up, and at higher speeds the rate of increase goes up." This description is more detailed, but still incomplete. There is more work to do to adequately characterize the relationship between stopping distance and vehicle speed. By plotting the data, you were able to grasp all data points at the same time. You could see that the relation between stopping distance and speed goes up faster for higher values of speed. In technical problem solving you call this step visualization of the data relationship.

"A picture is worth a thousand words" captures why making a simple data graph can be more revealing than gazing at a complicated data table for hours. This sets the background for the importance of this entire chapter.

6-2.1   Plotting Actual Numerical Data of Two Variables

The stopping distance/speed data in Table 6-1 is an example of actual numerical data. Assume that he data are from an actual experiment or observation, gathered in the context of the laboratory or as a result of field work or observation. The data typically will first be gathered in a data table like Table 6-1 where each row in the table represents a single set of values for all experimental variables, independent and dependent. In Table 6-1, the third row indicates a stopping distance of 128 feet was measured for a chosen value of 40 mph. for speed; that is one observation, or *data record*, from your experiment.

In making a plot of actual data, you must first determine what are the independent and dependent variables. We are going to limit the discussion in this chapter to two-dimensional plots, i.e., plotting only two variables at a time. If you want to deal with situations of more than two variables, you will need to make multiple data plots where each plot is of a dependent variable versus an independent variable. We will come back to this point below when we talk about subplots and overlay plots later in this chapter.

The practical reason for distinguishing between independent and dependent variables at the outset is that typically the independent variable is put on the *x*-axis and the dependent variable is put on the *y*-axis. Unless you are told otherwise, you should follow this rule.

In Section 6-1, you learned about the basic built-in functions you need to label a data plot:

1. **xlabel** to label the *x*-axis
2. **ylabel** to label the *y*-axis
3. **title** to create a title for the plot
4. **grid** to turn on (or off) an *xy*-grid

We will use the same labeling built-in functions for numerical plots.

To create the data plot shown in Figure 6-4 for the data in Table 6-1, the following MATLAB commands will do the trick. First, create two vectors to contain the data for the independent variable (**speed**) and the dependent variable (**stopDis**):

```
speed = 20:10:70;
stopDis = [46,75,128,201,292,385];
```

Note the use of the colon operator in the creation of the **speed** vector.

Now use **plot**, where before you used **ezplot**:

```
plot(speed, stopDis, '-ro')
```

The first argument to **plot** is the independent variable (that goes on the *x*-axis); the second argument is the dependent variable (that goes on the *y*-axis); we will explain the third argument later in this section. As always, the plot must be labeled.

plot can take multiple forms. In fact, **plot** is one of the most useful built-in functions in MATLAB, and one that you are sure to use many times. Open MATLAB Help and find Help for **plot**. The form used above is the third syntax form. Look under Description for the explanation of the third argument. In the paragraph on the third form for **plot**, you will see that the third argument is a *line spec* which means that you use the third argument to change the looks of the line plotted and the way the data points are plotted. In the paragraph, *line spec* is underlined, indicating a hot link to another description in MATLAB Help. Any time you see an underline like the one under *line spec* in MATLAB Help, click it for a further description of the underlined term, as you would if you saw an underlined hyperlinked term in your web browser.

The *line spec* we are using is **-ro**, indicating we want to connect the data points we are supplying to **plot** by a solid line (the **-** part of the *line spec*) that is colored red (the **r** part of the *line spec*) and to mark where each data point is with a small circle (the **o** part of the *line spec*). Click on the hot link in MATLAB Help that leads to the Help page for *line spec*. Read the page, and note the other options you have for different line types, colors, and data markers. Once completed, try out your knowledge by creating plots with the same data (**speed** and **stopDis**), but use the following:

- Blue lines that are dashed and with data markers that are asterisks
- Magenta lines that are dotted and with data markers that are diamonds
- No lines but with data markers that are five-pointed stars

Any time you plot real data, you need to include *data markers* for the data points in your data set. This is a firm rule in terms of the rest of this book, and for good reason. By including *data markers* in plots of real data, you indicate to anyone looking at the plot where the actual data points are in your plot and that your plot originates from real data. Both reasons are important.

Complete the plot by labeling it using the MATLAB commands below:

```
title('Stopping Distance versus Vehicle Speed, for
X2 System', ...
        'FontSize', 14)
xlabel('vehicle speed (mi/hr)', 'FontSize', 12)
ylabel('stopping distance (ft)', 'FontSize', 12)
grid on
```

The ellipsis (…) in the first line indicates to MATLAB a continuation of the line. The built-in functions **title**, **xlabel**, **ylabel**, and **grid** were introduced in the section above on **ezplot**. But now, more than one input argument exists. Find **title** in MATLAB Help. Look at the third syntactic form. This third form specifies that when you apply **title**, you can have any number of arguments, and then add a pair of new arguments to the arguments. The first of the pair is the *name of a Property* of the text that makes up the title, and the second of the pair is the *Property Value for that property*. **FontSize** is one property of text, whether the text is in the title of a plot or in the labeling for *x*-axis. Possible values of FontSize are 10, 12, 14, etc.

All of this sounds complicated, but it isn't. If you want to change the title, the *x*-axis labeling, or any text in MATLAB, then the function that creates the text (like **title**) takes optional pairs of **PropertyName** and **PropertyValue**. Almost all aspects of text may be set, but you need to know which **PropertyName** you will have to set to achieve your desired look for the text. **FontSize** is one of the most used text **PropertyNames**, as shown in the above code lines.

Using Help, determine what should be added to the **title** call above to make it so the title appears in **blue** font. The point is that few can remember all the property names for text in MATLAB. But seasoned MATLAB users do know where to find the list of property names.

You have more to learn, but you have the kernel now for making two-dimensional data plots using MATLAB.

6-2.2   Plotting functions using a numerical approach

In Section 6-1, you learned to use **ezplot** to plot a function. In this section, you will learn another, more flexible method: using **plot**. Many old hands at MATLAB prefer using **plot** over **ezplot**. You can see why if you look back at MATLAB Help for **plot** and compare it to the Help document for **ezplot**.

ezplot doesn't have the option of including <PropertyName, PropertyValue> pairs. This means that when you use **ezplot**, you do not have the comparable flexibility for controlling how plots will look once constructed.

In Section 6-2.1, the starting point was a data table, like Table 6-1. In this section, the starting point is a mathematical function. As a first example, consider the relation that gives the volume of a sphere given its diameter: $V(diam) = \frac{4}{3}\pi\left(\frac{diam}{2}\right)^3$. The goal is to generate a plot of this relationship using numerical plotting methods for values of *diam* between 1 and 100 meters. There are three distinct steps in creating a numerical plot of a mathematical relationship. First, you must create a data vector for the independent variable, which in this case is *diam*. Second, form a MATLAB expression based on the mathematical relation that will yield corresponding values of the dependent variable for each value of the independent variable. Third, generate the desired graph using the built-in function **plot** and the various built-in labeling functions. The MATLAB code below accomplishes all three steps:

```
diam = linspace(1,100);           % Step 1 - create vector
                                  for ind var
V = (4/3)*pi*(diam/2).^3;         % Step 2 - compute
                                  vector for dep var
plot(diam,V,'-r')                 % Step 3 - plot and label
xlabel('diameter (m)', 'FontSize', 12)
ylabel('Volume (m^3)', 'FontSize', 12)
title('Volume of a sphere versus its diameter','FontSize', 14)
grid on
```

Copy these commands to MATLAB and execute them.

There are several important observations about the general steps to create a plot of a function using numerical means and, specifically about the code above:

- Any method can be used to generate the vector of values for the independent variable, such as square bracket operator, colon operator, **linspace**, or any other method. Using **linspace**, as in the code above, is easy, particularly when 100 data values is enough to produce a good graph.

- When you create the vector of values for the dependent variable, you must take care to use DOT vector operations when necessary. The exponentiation operator to cube the radius (half the diameter) must be expressed as DOT exponentiation because the effect we want is to compute the cube for each of the data values in **diam**.

- The application of **plot** must generate a line but not generate data markers for the individual elements of the vectors for independent and dependent variables. If the third argument to **plot** in the example above is left out, a solid black line with no data markers will be generated. Data markers indicate the graph was generated using real data. The absence of data markers indicate that the data were generated from computed values for the dependent variable.

- The labeling (**xlabel**, **ylabel**, **title**) is necessary for any plot you create.

• Label the *y*-axis in the plot you create. MATLAB converted
  ^3 to a superscripted 3. This is one of the small features that
  make MATLAB plotting user-friendly.

Having worked through the steps to create a MATLAB plot of a mathematical
relationship using numerical means for a simple example, we will go on to a more
complex situation. The general steps described above remain the same. The
mathematical function $T(x,y)$ in Equation 6-1, is a function of two variables. Although
we are not going to do 3-D plotting in this chapter, we can still plot the relationship of
Equation 6-1 using numerical means. Look back to the section on **ezplot** (if you
need to) and think through why **ezplot** is not applicable to this situation. The
equation is:

$$T(x, y) = 92e^{-(x-1)^2}e^{-4(y-1)^2}$$

(EQ 6-1)

Physically, Equation 6-1 captures the situation sketched in Figure 6-5. A metal plate
of non-uniform material is located in the **xy**-plane as indicated; units for **x** and **y** are
meters. The plate is heated at the (1,1) corner by a constant heat source. Equation 6-1
describes the temperature T at any (x,y) point in the plate after enough time has passed
for the plate to come to equilibrium after heating has started. Units for **T** are degrees
Fahrenheit.

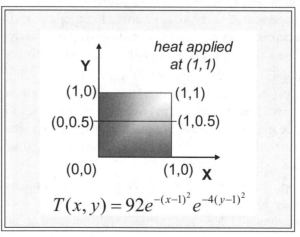

Figure 6-5: Metal Plate Heated at (1,1)

What is the temperature of the heat source? (*Hint*: Substitute in $x = 1, y = 1$ into
Equation 6-1 and examine the value of $T(1, 1)$. (Remember that $e^0 = 1$.) Although T

is a function of two variables, we can still apply **plot**. Suppose we limit values of **y** to be **y = 0.5**. This corresponds to the line shown in Figure 6-5 from (0, 0.5) to (1, 0.5). By holding **y** constant, at **y = 0.5**, Equation 6-1 is transformed to the following:

$$T(x)\big|_{y = 0.5} = 33.8e^{-(x-1)^2} \qquad \text{(EQ 6-2)}$$

We follow the same set of steps as used in the sphere volume problem above to generate a plot for the relationship of Equation 6-2. The plot resulting is shown in Figure 6-6:

```
x = linspace(0,1);          % Step 1 - vector for ind var
T = 33.8 * exp(-(x-1).^2);% Step 2 - vector for dep var
plot(x,T,'-r')              % Step 3 - plot and label
xlabel('x (m)', 'FontSize', 12)
ylabel('temperature for y=0.5 (deg F)', 'FontSize', 12)
title('plate temp, y=0.5: T = 33.8 exp(-(x-1)^2)','FontSize', 14)
grid on
```

Examine the plot in Figure 6-6. From the plot, the value of T at $x = 0$ (for $y = 0.5$) is about 12.5°F. Interpret this result from the graph in physical terms for the metal plate.

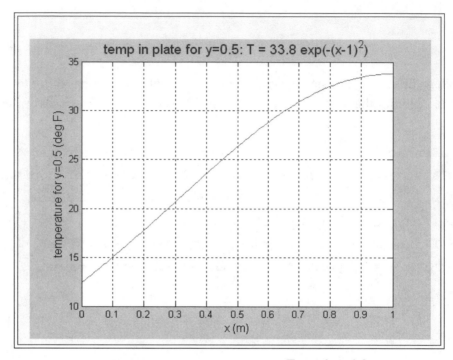

Figure 6-6: Numerical Plot for Equation 6-2

To help yourself bring together your knowledge of numerical plotting in MATLAB, do the following exercise.

> Write a MATLAB function that generates a graph like the one in Figure 6-6 for any value of y. Input to the function should be the value of y. There should be no output variables. (Don't worry for now about including the value of y in the title of the generated plot.) Test your function with values of y in {1, 0.5, 0}.

1. **Use `plot`, a kernel built-in function, for creating graphs of numerical data.**

2. **The first argument to `plot` should be the vector of values for the independent variable (along the x-axis); the second argument should be the vector of values for the dependent variable (along the y-axis).**

3. **An optional third argument `plot` is the line spec which specifies the type of line used (solid, dotted, etc.), the color of the line used, and the type of data marker (if any).**

4. **An optional pair of following arguments express properties of the line.**

5. **Use `figure` to create a new plotting window.**

6. **`xlabel`, `ylabel`, and `title` are used for labeling a plot.**

7. **`grid on` will activate a grid in the `xy`-plane of the plot. `grid off` will deactivate the grid.**

8. **For plotting numerical data from experimentation or observation, data markers are mandatory.**

9. **For plotting numerical data that are computed from a mathematical relationship, data markers must not be used.**

Synopsis for Section 6-2

6-3   Overlay Plots and Subplots

You have the rudiments now for constructing meaningful plots of mathematical relationships (using **ezplot** or **plot**) and of data from experiment or observation (using **plot**). However, you have seen situations in which we plotted only one mathematical relationship or experimental data relationship in a given plot window. MATLAB allows extension beyond that; the extensions include overlay plots, subplots, using multiple y-axes (**plotyy**), and others. Each method has its own functionality, and its own criteria for when it may best be used. In this section, we consider *overlay plots* and *subplots*.

6-3.1   Overlay Plots

The purpose of an overlay plot is to allow putting more than one relationship directly into the same plotting window. All of the built-in functions you have learned still apply with two new built-in functions to learn: **hold** and **legend**.

Return to Equation 6-1 that captured the variations of temperature on a heated metal plate. We transformed Equation 6-1 to Equation 6-2 by fixing the value of y to be $y = 0.5$ and, thus, enabled making a plot of $T(x)|_{y = 0.5}$ for x in $\{0,1\}$. If we could construct a number of such plots for carrying values of y and "glue them" into the same window, we would have a family of curves for T which would give us a good picture of $T(x, y)$ over the whole metal plate. The MATLAB code below will produce a plot with such a family of relations shown for $T(x, y)$, and the resultant plot is like that shown in Figure 6-7.

```
1    xV = linspace(0,1);  % vector of x values
2    yV = 0:0.25:1;        % vector of y values that will be used

   % temperature distribution vectors
3    temp5 = 92*exp(-(xV-1).^2)*exp(-4*(yV(5)-1)^2); % y=1.00
4    temp4 = 92*exp(-(xV-1).^2)*exp(-4*(yV(4)-1)^2); % y=0.75
5    temp3 = 92*exp(-(xV-1).^2)*exp(-4*(yV(3)-1)^2); % y=0.50
6    temp2 = 92*exp(-(xV-1).^2)*exp(-4*(yV(2)-1)^2); % y=0.25
7    temp1 = 92*exp(-(xV-1).^2)*exp(-4*(yV(1)-1)^2); % y=0.00

   % make the chart
8    figure, hold on
9    plot(xV, temp5, 'r', 'LineWidth', 2) % y=1.00
10   plot(xV, temp4, 'c', 'LineWidth', 2) % y=0.75
11   plot(xV, temp3, 'g', 'LineWidth', 2) % y=0.50
12   plot(xV, temp2, 'b', 'LineWidth', 2) % y=0.25
```

```
13   plot(xV, temp1, 'k', 'LineWidth', 2) % y=0.00
14   legend('y=1.00 m','y=0.75 m','y=0.50 m','y=0.25 m','y=0.00 m', 2)
15   xlabel('x (m)', 'FontSize', 12)
16   ylabel('Temp (deg F)', 'FontSize', 12)
17   title('Temp vs Location for Heated Metal Plate', 'FontSize', 14)
18   grid on
19   equationTxt = 'T(x,y)=92 * exp(-(x-1)^2) * exp(-4 * (y-1)^2)';
20   text(.3,95, equationTxt, 'FontWeight', 'Bold')
```

The numbers on the left in the code above are line numbers, like the line numbers if this code were in the MATLAB editor window. Put up this code in MATLAB and run it. You should get a plot up like that shown in Figure 6-7.

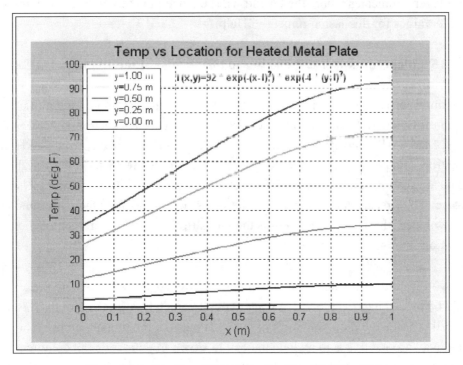

Figure 6-7: Family of Solutions for Heated Metal Plate

Line 1 is an application of **linspace** to set up the major independent variable for the problem, a vector whose elements are positions on the *x*-axis. **Line 2** is setting up the other independent variable, a vector whose elements are positions on the *y*-axis. Our path is to set up the independent variable **xV**, and for each one of the values in **yV**, find the corresponding $T(x)|_y$ vector. **Lines 3-7** build the five vectors we seek.

Line 8 creates a new blank plotting window and turns `hold on` so more than one plot can appear on the window. Without `hold on`, the second call to `plot` would wipe out the first, the third would wipe out the second, etc. **Lines 9-13** create the plot lines for each of the five values of `y`. Note the use of `LineWidth`, a property that can be set for any line.

Line 14 is a call to the built-in function `legend`. As its name suggests, `legend` creates a legend for the plot. The first five arguments are *strings*, set off by single quotation marks. These are descriptive phrases for the five overlay plots we have put into this window. The order of the strings must be the same as the order in which the plot lines were created. This ensures the color of each of the plots will be written with its intended descriptive phrase. If there were seven overlay plots put in this window, then we would create a legend with seven descriptive strings. The whole purpose of the legend is to help us to distinguish the lines in the chart. The last argument in the call to legend indicates where the legend will be placed in the window. Look up the possible values for this last argument in Help.

Lines 15-17 label the *x*-axis and *y*-axis, and title the plot. **Line 18** turns on the *xy*-grid.

Many times, especially for more complex plots, labeling axes and giving a title is not enough information. In such cases where further annotation is desired, several options exist in MATLAB. One of the easiest is the built-in function `text`, which takes an annotation, and places it at a specified point in the plot. **Line 18** sets up a variable to hold an annotation: a description of the mathematical relation $T(x,y)$. **Line 19** is a call to `text`. The first argument to `text` is the *x*-position in the plot where the annotation is to be placed; the second argument is the *y*-position; and third argument is the annotation.

But if you want to annotate a plot by using text, how do you know where the text should be placed into the plot? You want to put it somewhere on the plot that is empty. But until you see the plot, you don't know what parts of the plot are empty. The typical way around this is to include the call to `text` in your code but comment it out until you see a sample of the plot. Then you can go back, put in the position where you want the annotation placed, and uncomment the line of code.[6] Another way to annotate a plot is by the use of `gtext`. Use Help to look up what the functionality of `gtext` is, and how you use `gtext`.

---

6. Making a line of code a comment is effected in the Edit window by placing the cursor in the target line, right-clicking, and selecting "Comment." Likewise, changing a line of code from a comment to a command is accomplished in the same way, but as a last step, you select "Uncomment."

Be sure you are clear on what overlay plots are and under what circumstances you would want to create an overlay plot.

6-3.2   Subplots

The selection of an overlay plot is most often made to display a family of parametric plots all in the same chart window, as shown in Figure 6-7. For a function of two variables $f(x, y)$ we create a set of functions $f(x)|_y$, meaning we pick a value of y to use for each member in the family. Then we plot all these versions of $f(x)|_y$ to the same chart window.

Suppose we want to show the behavior of a set of dependent variables of a system but that these dependent variables hold data that are not of like type. Suppose also that we want to show in the same window the two relations that describe the velocity and the acceleration of an object as a function of time. The clue that we should not think about using an overlay plot is that velocity and acceleration have different units; therefore, we cannot legitimately make an overlay plot using one y-axis for both. Such a problem, however, is classic for the use of a set of subplots.

Subplots are best thought of as independent plots, all put into the same graphics window. Consider each of the independent plots as being placed into a pane of the graphics window. Constructing the whole window then becomes an exercise in telling MATLAB the location of a given pane, then using built-in functions like **plot** to construct the contents for that pane. You know how to use **plot** to construct an individual plot. The key new item to learn is how to tell MATLAB where panes are located in a graphics window. The MATLAB built-in function that supports placement of window panes is **subplot**. Look up **subplot** in Help. The starting point for understanding **subplot** is to think of a graphics window as a table, such as an Excel table. We term such a table an **n-by-m table** (most often written $n \times m$) to mean that it has n rows and m columns. A graphics window in MATLAB can be laid out (the technical term is gridded) to have n rows and m columns.

subplot takes three input arguments: **subplot(n,m,paneNumber)**. The first two (n and m) are the number of rows and columns in the grid to be placed on the graphics window. The third argument is the pane number. MATLAB numbers the panes in a graphics window starting with the first row, and at the left. The numbering scheme and the calls to the **subplot** built-in function that would be used to tell MATLAB that we want to put a plot into a given pane are shown in Figure 6-8 for an example situation.

subplot(2,3,1)	subplot(2,3,2)	subplot(2,3,3)
PANE 1	PANE 2	PANE 3
subplot(2,3,4)	subplot(2,3,5)	subplot(2,3,6)
PANE 4	PANE 5	PANE 6

Figure 6-8: Numbering of Window Pane in a Graphics Window and Call to Subplot to Identify a Given Window Pane

Notice in the Help description of the first form of **subplot** that the third argument can be a vector. If the third argument is a scalar, then it indicates directly the pane to which plots are directed. If the third argument is a vector, then any plot is directed to cover all numbered panes indicated in the vector. For example, **subplot(2,3, [1,4])** in relation to Figure 6-8 would mean that any plot would be directed to cover the panes number 1 and 4. The ability to cover multiple panes in this way is useful for spacing of x-labeling so one subplot pane does not interfere with another. Using a vector for the third argument of **subplot** is shown in the example below.

To illustrate the use of subplot, let us use a simple situation: an object in free space that starts at time = 0 at some location X_0, with initial velocity V_0. Assume a constant acceleration A in the same direction as the initial velocity. In this situation, the acceleration as a function of time, the velocity as a function of time, and the location as a function of time are the following equations:

$$a(t) = A$$
$$v(t) = V_0 + At$$
$$x(t) = X_0 + V_0 t + \frac{1}{2}At^2$$

(EQ 6-3)

Suppose you want to create a graphics window that shows all three plots for a given initial position and velocity, and given constant acceleration, where time runs from {0,100} seconds.

The first thing you must do is to determine how you want to *tile the window*, that is, how you want to lay out the window panes in the graphics window you create. Suppose you want to show the three graphs as shown in Figure 6-9. The MATLAB code below will accomplish your plotting goal.

```
function plotFreeBodyMotion(initPos, initSpeed, constAccel, T)
% plot free body motion with constant acceleration for time [0, T]
% INPUTS: initPos - the initial position in some coordinate system (m)
%         initSpeed - the initial speed (m/s)
%         constAccel - the constant acceleration of the object (m/s^2)
%         T -  the end point time (s)
% OUTPUTS: none

  time = linspace(0,T);                               % s
  accel = constAccel * ones(1,length(time));          % m/s^2
  vel = initSpeed + constAccel * time;                % m/s
  pos = initPos + initSpeed*time + 0.5*constAccel*time.^2;% m

% plot the results
  fStr = 'init pos=%3.0f (m)...init vel=%3.0f (m/s)';
  desctiptiveStr = sprintf(fStr,initPos, initSpeed);

  figure
   subplot(30,1,[1,5])
    plot(time, accel, 'g', 'LineWidth', 2)
    ylabel('acceleration (m/s^2)')
    title('Acceleration vs Time (sec)', 'FontWeight', 'bold')
    text(10,3.5,desctiptiveStr)

   subplot(30,1,[10,15])
    plot(time, vel, 'b', 'LineWidth', 2)
    ylabel('speed (m/s)'), grid
    title('Speed vs Time (sec)', 'FontWeight', 'bold')

   subplot(30,1,[20,30])
    plot(time, pos, 'r', 'LineWidth', 2)
    ylabel('position (m)'), grid
    title('Position vs Time (sec)', 'FontWeight', 'bold')
```

Put this code up as a function in MATLAB. Notice the use of **subplot** with the third argument being a vector.

The variables **fStr** and **descriptiveStr** are set up, and a text formatting function (**sprintf**) creates the desired annotation strings. Look up **sprintf** in Help. The string **descriptiveStr** is used to help properly label the graphics window that is constructed. It is given to **text** for placement in the topmost subpane. Call **plotFreeBodyMotion** with the following command:

```
plotFreeBodyMotion(52,35,3,100)
```

The graphics window you should generate is shown in Figure 6-9. Using a vector for the third argument to **subplot** allows you more control over how you space the multiple plots you want to display. To see why this makes for more tidy graphics windows, go back to the function **plotFreeBodyMotion**, and change **subplot** to number three panes (**subplot(3,1,N)**) and compare the different effect.

The mechanics of using **subplot** are not difficult once you become familiar with them. The key to the mechanics is the tiling of the graphics window as shown in Figure 6-8. The ability to make a multiple plot window like that in Figure 6-9 is useful in many different contexts.

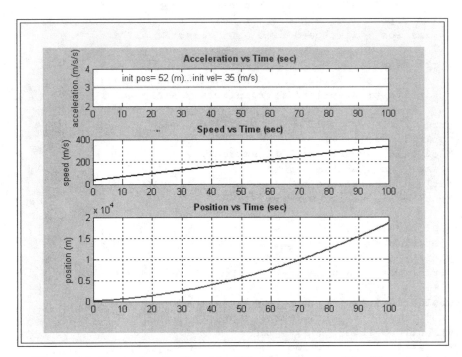

Figure 6-9: Example of a Graphics Window Constructed with subplot

6-3.3   Section Synopsis

In this section, you have learned to create subplots. In addition to learning the mechanics in MATLAB for subplot generation, you need to remember under what circumstances a set of subplots is appropriate. When you want to show a number of results for different dependent variables graphically, then using **subplot** makes sense. This is in contrast to the appropriate use of overlay plots when you have a family of results for the same independent variable, and you want to best show the relationships for the entire results family.

---

1. **Overlay plots are used to show a family of parameterized results.**
2. `hold on` **is the key MATLAB command needed to turn on overlays.**
3. **Subplots are used to display plots of different independent variables, usually from one experimental data set or from one set of equations for a single physical system.**
4. `subplot` **is the key MATLAB command needed to identify the target for a created plot.**

Synopsis for Section 6-3

6-4   3-D Plotting with MATLAB

Let's go back to basics for a minute. What is the *purpose* for making any plot? Although you have gone through a fast and detailed introduction to 2-D plotting with MATLAB, you need to remember why you are learning the capabilities presented in this chapter.

The purpose is to *communicate*. If you are using `ezplot` to understand a given functional relationship, then you want to communicate to yourself. If you are using `plot` to summarize an experimental dataset between an independent variable (speed of car) and a dependent variable (amount of damage in a crash), then you may be "communicating with yourself" or you may be developing something to communicate to your boss. In any case, the important purpose of developing a graph is to communicate in a concise, accurate manner.

In this section, you will extend your ability to communicate with MATLAB plots to include three-dimensional plots. Earlier in this chapter, you studied `ezplot` and `plot`. Both can be used exclusively for 2-D graphs, although in Section 6-3 you learned how to use overlay plots and subplots to show more than one 2-D graph in one window. Still, you are fundamentally limited when using `plot` and `ezplot` to making 2-D plots.

In the rest of this section, you will learn to create 3-D plots, both functional plots and data plots.

6-4.1   Creating 3-D Functional Plots (`ezplot3` and `ezsurf`)

The 2-D plotting function **ezplot** can be generalized to 3-D plotting in two ways. First, you can create a 3-D plot that is a graph of three functions, each of the same (one) independent variable. This is the analog of using **ezplot** to create a 2-D plot that shows two functions each of the same (one) independent variable. This was the parameterized version of **ezplot**. The analogous 3-D built in function is **ezplot3**.

The second generalization of **ezplot** comes into play when you want to create a graph of a single function of two independent variables. The first form of **ezplot** you learned created a 2-D graph of a single function of one independent variable. A number of 3-D plotting function are used to plot a function $f(x,y)$; three of them are **ezsurf**, **ezmesh**, and **ezcontour**.

6-4.1.1   Using `ezplot3` to Create Parameterized Line Plots in 3 Dimensions

The third form of **ezplot** was the parameterized form and was used when you wanted to create a functional plot for two functions, both having the same independent variable, e.g., $f(t) = sin(t), g(t)=cos(10t)$, for $0 < t < \pi$. You learned that such a parameterized plot can be created by

```
ezplot('sin(t)', 'cos(10*t)', [0, pi])
```

Remember the type of question such a parameterized plot can be used to answer directly, e.g., the value of $sin(t)$ when the value of $cos(10t)$ is 0.5.

Now suppose you have three functions: $f(t)$, $g(t)$, and $h(t)$. All three have the same independent variable t, and you want to create a parameterized plot that will show a three-space curve of connected points such that for a given value of t

- the value of $f(t)$ is on the x-axis,
- the value of $g(t)$ is on the y-axis, and
- the value of $h(t)$ is on the z-axis.

To create a plot like this, the function you will want to apply is **ezplot3**. To make the situation concrete, suppose $f(t)=sin(t)$, $g(t)=cos(t)$, and $h(t)=t$. Additionally, you want t to range over [0, 20]. The call to you need to make is as follows:

```
ezplot3('sin(t)', 'cos(t)', 't', [0,20])
```

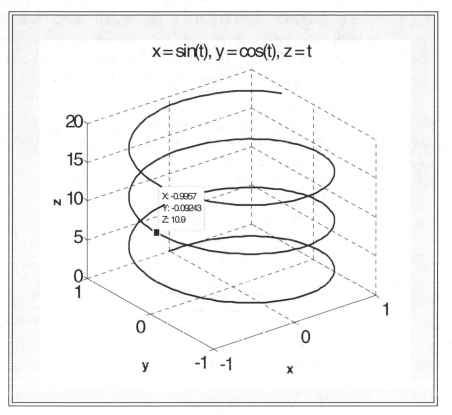

Figure 6-10: Example of graph made with `plot3`

The figure you produce will look like Figure 6-10.

Although the code above will not produce it, note the highlight for $t = 10.9$. The highlight was produced using the data cursor function. By clicking on the icon shown in Figure 6-11 for creating a data cursor, you can label any point on a graph (2-D or 3-D). You should consult MATLAB HELP for details about using the data cursor tool.

Figure 6-11: View of the figure toolbar

To help you understand a 3-D plot, in MATLAB you can rotate the three-dimensional graph in its 2-D projection on your computer screen. To "turn on" the rotate function, click on the icon indicated in Figure 6-11 in the toolbar. Rotation of a 3-D projection can often help in understanding a 3-D functional or a data plot.

For example, in Figure 6-12, the plot first shown in Figure 6-10 is rotated. After rotation, you can see that the line being tracked by $f(t)$, $g(t)$, $h(t)$ is a spiral path that could be described as "moving up a cylinder."

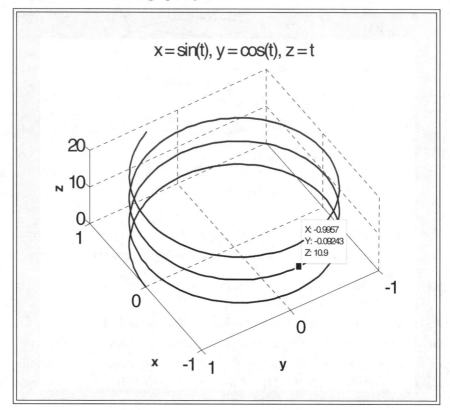

Figure 6-12: A rotated view of the plot first shown in Figure 6-10

Create the plot indicated for yourself and test the "rotate" capability. Also, test the "animate" option for **ezplot3** and consult HELP.

6-4.1.2   **ezsurf** to Create Surface Plots for Functions of Two Variables

As you plotted a function of one variable using **ezplot**, MATLAB allows you to create a plot of two variables, $f(x,y)$, using **ezsurf**.[7] Suppose we return to the

---

7. Other functions also plot a function $f(x,y)$, **ezsurf** and **ezcontour** to name two.

equation for heat transfer in a block that we used in Section 6-3. That equation (reproduced here from Equation 6-1) is as follows:

$$T(x, y) = 92e^{-(x-1)^2}e^{-4(y-1)^2}$$ **(EQ 6-4)**

This equation represents the temperature profile across a plate of a given material as
shown in Figure 6-5. When we analyzed this relation earlier, we used a number of 2-D
plots, both overlay plots and subplots, to develop a family of graphs based on holding
x (or y) to one fixed value and on allowing the other variable to change. With 3-D
plotting capability, we can create one graph that will show the entire story and do it
with one MATLAB command. MATLAB code to create a 3-D plot of the two-
dimensional function $T(x,y)$ using the built-in function **ezsurf** is as follows:

```
ezsurf('92*exp(-(x-1)^2)*exp(-4*(y-1)^2)', [0,1,0,1])
```

When you execute this command, you will get a 3-D graph similar to the one in
Figure 6-13. Note that the labels and title are generated automatically.

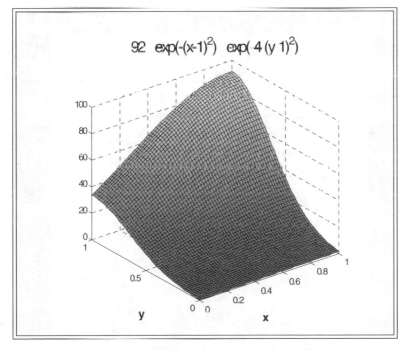

**Figure 6-13: Surface plot of temperature profile in a square
plate with edge 1, heat source at (1,1)**

Other 3-D functional plotting forms are available as well. For example, to make a wire frame plot rather than a surface plot for a function of two variables, the built-in **ezmesh** is available. To make a contour map for a function of two variables, the built-in **ezcontour** is available.

We will do an example of one more 3-D ez-form to end this chapter: **ezcontour**. You see a contour plot when you watch the evening news and see the weather person show a weather map with lines of equal barometric pressure. Although a contour plot appears to be 2-D, it is showing a 3-D function or dataset.

For the functional plotting form (the ez-form), we start with a function of two variables, *f(x,y),* as we did for **ezsurf**. Instead of visualizing *f(x,y)* as a height above the *xy*-plane, we will visualize it by connecting points in the *xy*-plane that produces the same *f(x,y)* value. The *xy*-plane is, thus, divided into regions that show how *f(x,y)* varies.

For example, returning to the relation of Equation 6-4, the following code will produce the following plot:

```
ezcontour('92*exp(-(x-1)^2)*exp(-4*(y-1)^2)', [0,1,0,1])
```

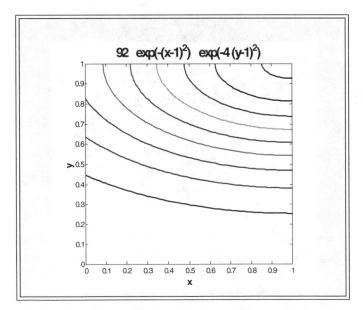

Figure 6-14: Contour plot of temperature profile in a square plate with edge 1, heat source at (1,1)

In Figure 6-14, the contour lines "bunch" higher on the y-axis than they do on the x-axis. This reflects the multiplier (4) on the y term in the relationship compared to the multiplier on the x term (1).

A variant on contour plotting is the filled contour plot type. Try the following:

```
ezcontourf('92*exp(-(x-1)^2)*exp(-4*(y-1)^2)',[0,1,0,1])
```

Consult HELP to browse the entire range of possibilities.

6-4.2   Creating Data Plots for Datasets of Three Variables Using `meshgrid`, `griddata`, and `mesh`

In the last section, you saw some of the capability of MATLAB for creating functional plots using `ez-form` plotting built-in functions. You can create 3-D functional graphs using the built-in function `mesh` just as you could create a 2-D functional graph using `plot`. (See Section 6-2.) In this section, the focus will be on creating 3-D data graphs for a dataset of three variables. The 2-D analog is making a 2-D scatter plot for a dataset of two variables and connecting the data points to help visualize a relationship.

Suppose we start with a dataset of three variables: x, y, z. Data included in the dataset are shown in Table 6-2.

Table 6-2: A dataset containing three variables: x, y, z

x	y	z
-1.12	0.12	-0.31
-1.81	0.68	-0.04
0.72	-1.97	0.01
0.73	-0.47	0.34
1.74	-1.73	0.01
1.13	0.52	0.25
-0.47	-0.33	-0.34
0.08	0.75	0.04
1.32	0.36	0.20
0.01	0.02	0.19
-1.79	1.39	-0.03

The steps that follow can be used when you have values for two independent variables. In Table 6-2, that would be x and y. You suspect a third variable, z, that you think of as a function of the other two, i.e., $z=f(x,y)$.

For an explicit example, suppose you have measured the speed of a crash test car, you know its weight, and you expect the stress on a test dummy to be a function of car speed and weight.

To analyze the data in Table 6-2, you can use the following MATLAB code:

```
1  rangeX = linspace(min(x), max(x), 20);
2  rangeY = linspace(min(y), max(y), 20);
3  [XI,YI] = meshgrid(rangeX ,rangeY );
4  ZI = griddata(x,y,z,XI,YI);

5  figure
6  mesh(XI,YI,ZI), hold on

7  plot3(x,y,z,'rp', 'MarkerFaceColor', 'r', ...
                'MarkerSize', 10)
8  xlabel('x', 'FontWeight', 'bold')
9  ylabel('y', 'FontWeight', 'bold')
10 zlabel('z', 'FontWeight', 'bold')
11 title('Mesh Plot:Dataset of Three Variables', ...
          'FontSize', 16)
12 colormap('winter')
```

This will produce the information Figure 6-15.

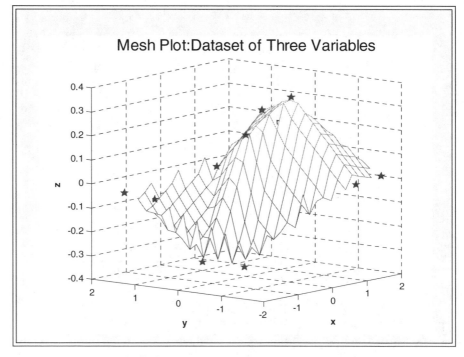

Figure 6-15: 3-D mesh plot of a dataset of three variables

Enter the previous lines of code into a MATLAB script file and run the script. Once you produce the graph (similar to Figure 6-15), turn on the ROTATE function and "play" with various orientations of the axes to understand the mesh surface you have created.

Here is additional information of lines 1–3:

- **Lines 1-2** in the code above find the ranges on values for x and y and use **linspace** to lay out a uniformly spaced vector covering the range of x and y.

- **Line 3** uses a built-in function you have not seen before: **meshgrid**. The purpose of **meshgrid** is to lay out a uniform grid that spans a given range along the x and y axes. We have the ranges on x and y from **Lines 1-2**.

To understand what **meshgrid** produces, execute the following command and note the result:

```
>> [gridX, gridY] = meshgrid(-1:1, -10:10:10)
gridX =
          -1.00              0              1.00
          -1.00              0              1.00
          -1.00              0              1.00

gridY =
         -10.00         -10.00         -10.00
              0              0              0
          10.00          10.00          10.00
```

We input two arguments to **meshgrid**: a vector uniformly covering a range of x values and a vector uniformly covering a range of y values. We were handed back the coordinates of a grid that uniformly covered the entire xy plane that we specified. *gridX* and *gridY* are *arrays*, the topic of Chapter 7 in this textbook. For now, think of two tables with three rows and three columns. Corresponding elements of the two tables together form an *XY* point on a grid.

Examine Figure 6-16. Each of the solid circles is at a location specified by the *XY* points in the combination of *gridX* and *gridY*. Be sure you make the mental connection between the data in *gridX* and *gridY* and the diagram shown in Figure 6-16.

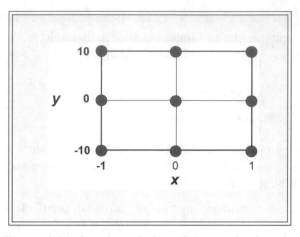

Figure 6-16: Conceptual view of the result of meshgrid

Now that you understand **`meshgrid`** in **Line 3**, we can pick up the story again at **Line 4**.

- **Line 4** is a call to a second built-in function you have not seen before: **`griddata`**.

The variable values in the dataset we started with were non-uniformly distributed across the range of *x* and *y*. To use the standard graphing methods to make a 3-D data plot, we will need a uniform coverage of the ranges on *x* and *y*, like the one in Figure 6-16.

But suppose the "real data" falls in the *xy* plane as shown in Figure 6-17.

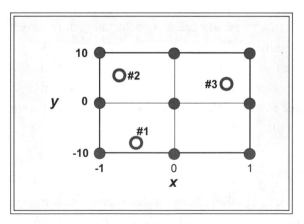

Figure 6-17: Sample of "real data" superimposed on a mesh

The open circles in Figure 6-17, marked #1, #2, and #3, are xy values from a non-uniformly distributed dataset. Remember, above each of these points is a corresponding z-value. What we need to do is "interpolate" from the z values for points #1, #2, and #3 to get an "inferred" z value above each point in the grid. The call to **griddata** in **Line 4** provides what we need.[8]

The first three input arguments of **griddata** are vectors containing the data elements for the three variables we want to plot. The last two input arguments are the uniformly distributed arrays that define an XY grid analogous to the one shown in Figure 6-16. The single output variable is a Z array that gives a z value for every corresponding *vertex* in the data grid.

Now, we can return to the code:

- **Line 5** creates a new plotting window.

- **Line 6** is a call to the function **mesh**; **mesh** creates a "wire frame" 3-D plot for a function of two variables $f(x,y)$. The interpretation is that every vertex on the xy grid (produced by **meshgrid**) has a z value above it (produced by **griddata**), and all of these z values define the surface graphically displayed by the wireframe.

- **Line 7** is a call to **plot3**, which is used for plotting a 3-D line defined by three functions of the same independent variable. **plot3** is analogous to **ezplot3**, the difference being that **plot3** is used for plotting real data. This call to **plot3** is used for creating the data markers shown in Figure 6-15.

- **Lines 8-11** are used for labeling. You have seen these functions before.

- **Line 12** is new: **colormap** is a built-in function used to set the range of colors used in any plot, in particular in the wire frame created by **mesh**. Look up **colormap** in HELP for the names of commonly used maps.

To create a 3-D data plot, the two built-in functions you need to master are **meshgrid** and **griddata**.

---

8. Google "interpolation" if you want to refresh your memory on this operation. The specifics for interpolation methods available in MATLAB are, of course, in HELP.

6-4.3   Creating Functional Plots for Functions of Two Variables Using `meshgrid`, and `surf`

In Section 6-2.2, you learned how to create graphs of functions of one variable using **plot**. The basic story was to create a vector for the independent variable typically by relying on **linspace**, to compute corresponding values of the dependent variable by forming a vector expression, and to create the graph desired by calling **plot**.

The analogous operation in 3-D is used to create a graph of a function of two variables, $f(x,y)$. It uses **meshgrid** exactly as it was used in the last section to create a regularly spaced grid in the XY plane but does not use **griddata** because interpolation from actual data to a grid is not needed. No "real data" are involved in this case. The call to **griddata** is replaced by computing $z = f(x,y)$ values using an *array* expression for every xy vertex produced by **meshgrid**. Although you will not examine arrays and array expressions until the next chapter, the cell-by-cell array operations we need here are directly analogous to vector cell-by-cell operations used to form a vector expression.

Suppose we want to graph this function:

$$f(x, y) = 10\, e^{-2y}\, \text{sech}(5.2x)$$

(EQ 6-5)

The following MATLAB code, for example, creates the functional plot in a manner that is analogous to the way you learned to use **plot** in Section 6-2 to develop 2-D functional plots in a numerical manner.

```
1  x = linspace(-1,1, 20);
2  y = linspace(-1,1, 20);
3  [X,Y] = meshgrid(x,y);
4  Z = 10 * exp(-2*Y) .* sech(5.2*X);
5  figure
6  surf(X,Y,Z)
7  title('f(x,y) = 10 e^-^2^y sech(5.2 x)', ...
       'FontSize', 16, 'FontWeight', 'bold')
8  xlabel('x', 'FontWeight', 'bold')
9  ylabel('y', 'FontWeight', 'bold')
10 zlabel('f(x,y)', 'FontWeight', 'bold')
```

Enter this code into a script in MATLAB and execute it. The graph you produce should be similar to that below although Figure 6-18 has been "rotated."

Now, let's analyze the code.

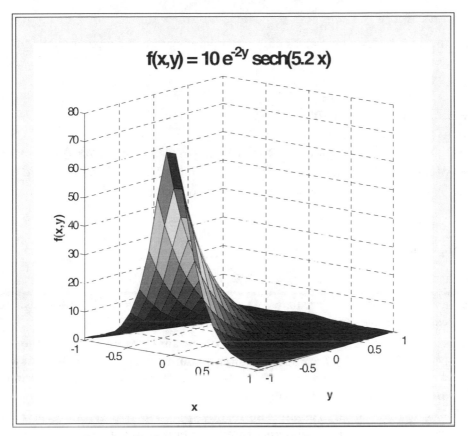

$$f(x,y) = 10\,e^{-2y}\,\mathrm{sech}(5.2\,x)$$

Figure 6-18: 3-D mesh plot for a function of two variables

- **Lines 1-2** set up uniformly-spaced vectors for the range of x and y that we use as the domain of $f(x,y)$.

- **Line 3** uses **meshgrid** as you learned previously for developing a 3-D data plot. Arrays X and Y are produced.

- **Line 4** is an *array* expression that mirrors the relation in Equation 6-5.

- **Line 5** produces a new graphics window.

- **Line 6** is a call to **surf**, which produces a 3-D surface map. Either **surf** or **mesh** could be used; the former producing a surface plot and the latter producing a wire frame plot.

- **Lines 7-10** are labeling operations for the graph.

One more built-in function can be useful when creating surface plots. Often it is helpful to build the surface so it is partially transparent. The built-in function **alpha** sets the transparency of any graphic plot but is particularly useful for surfaces. After you enter the previous code and produce the surface plot similar to that of Equation 6-18, in the command window, execute the following:

```
alpha(0.8)
```

Notice the effect. Try other values of the transparency (the one input argument for alpha).

6-4.4   Section Synopsis

In Section 6-4, you have learned to create 3-D plots using MATLAB. A top level group of the plot types was on the symbolic functional plot for which you used ez-forms. A second group of plot types was capable of producing both data and functional plots.

A synopsis for Section 6-4 follows:

---

1. `ezplot3` is used to create a 3-D plot when you have three functions, each of the same independent variable: *f(t), g(t), h(t)*.

2. `ezmesh` can be used to create a wire frame surface plot for a function of two variables: *f(x,y)*. `ezsurf` creates a similar graph, but instead of a wire frame, it produces a surface plot. `ezcontour` produces a contour plot for a function of two variables.

3. `meshgrid` is used to create a uniformly spaced grid of *xy* values, given a uniform vector of *x* values and a uniform vector of *y* values.

4. `griddata` is used to interpolate a (possibly) non-uniform dataset to create a uniform dataset suitable for the major 3-D surface plotting methods.

5. Once a uniform dataset is created, `surf`, `mesh`, and `contour` can all be applied.

6. `colormap` is used for changing the range of colors used in any graph and is particularly useful for surface plots.

7. `alpha` is used to set the transparency of any graph and is particularly useful for surface plots.

Synopsis for Section 6-4

6-5   Other Plot Types and Other Tools

In this chapter you have worked with a number of plot types, both 2-D and 3-D. With you have already mastered in this chapter, you have a good working knowledge in creating both 2-D and 3-D graphs. But what you have learned so far is only the tip of the iceberg.

To get an idea of the MATLAB plotting types, go to Help. Click on the tab labeled **Demos**, then **MATLAB**, and then **2-D Plots** or **3-D Plots** or **3-D Surface Plots**. In the Help information pane, click on the blue underlined **Run this demo**. The demos in Help have many interesting applications of MATLAB, and is a good place for anyone wanting to know general capabilities of MATLAB.

If you want to explore yet further, click on the Contents tab in Help. Then click on **MATLAB, Graphics,** and **Creating Specialized Plots**. The Contents part of Help are an organized listing of MATLAB topics amounting to a relatively fast online textbook for use of MATLAB capabilities.

In addition, MATLAB offers a number of interactive tools to allow you to get "down and dirty" with your datasets. For one, MATLAB has a very robust, interacttive utility to help you do *function finding/function fitting*. Many times in engineering problem solving, you have a dataset that you would like to characterize with a mathematical function. You have seen in this chapter how to create data plots to help you visualize relationships that exist in the data. But MATLAB goes substantially beyond that with a tool called the *Basic Fitting Tool*. This tool is available in any graphical plot window. Its use is explored in Problem 6-B.16, Problem 6-B.17, Problem 6-B.19, and Problem 6-B.20.

For a second, MATLAB includes another interactive tool that can be invoked with the built-in function `figurepalette`. This tool is useful esepcially for interactively creating subplots and overlay plots, and in general for graphically exploring a number of variables in your top level workspace. The use of `figurepalette` is explored in Problem 6-B.18.

6-6   Problem Sets for 2-D and 3-D Plotting

In Sets A and B below, put your solution in the form of a MATLAB function unless you are told to do otherwise in a problem spec.

Don't forget to label (including units when available) your plotting figures fully.

6-6.1   Set A: Nuts and Bolts Problems for 2-D and 3-D Plotting

In Set A, problem specs are set so you are asked for some of the problems to save the figure you create. At other times, you will be asked to create a function that creates a graph and save the function. Pay attention to what you are asked to do in any given problem.

Problem 6-A.1 (Section 6-1)

Use **ezplot** to plot the following function in the closed interval $[-2\pi, 2\pi]$:

$$f(x) = \sin(x^2)$$

Write a function that creates this plot. Name the function **Prob6_A_1**.

Problem 6-A.2 (Section 6-1)

Use **ezplot** to plot the following function in the closed interval $[0, A\pi]$:

$$f(x) = \sin(x)e^{-0.1x}$$

A is the parameter that you set.

Write a function that creates this plot. Your function will take one input which is a value for A. Name the function **Prob6_A_2**.

Call your function with $A = 2$. Write a short paragraph describing the plot you produced.

Call your function with $A = 8$. Write a short paragraph describing the second plot you produced.

What lesson is to be learned from this problem?

Problem 6-A.3 (Section 6-1)

Use **ezplot** to plot the following polynomial and graphically determine its two roots:

$$f(x) = -2x^2 + 5x + 4$$

Turn the grid on and enlarge the graph around the roots to increase accuracy. Search in MATLAB Help to learn how to zoom a 2-D graph.

Label the roots in the plot using **text** or **gtext**. Save the figure and call it **Prob6_A_3.fig**.

Problem 6-A.4 (Section 6-1)

Use **ezplot** to determine the roots graphically of the following equation in the interval $[0, 2\pi]$:

$$x\tan(x) = 9$$

Use the grid and zoom facilities of MATLAB for more accurate answers.

Label the roots in the plot using **text** or **gtext**. Save the figure and call it **Prob6_A_4.fig**.

Problem 6-A.5 (Section 6-1)

Use **ezplot** to plot the following implicit function in the interval $[-6, 6]$:

$$f(x, y) = 0 = (-2x^2y^2) + 5x^2y - 4xy^2 + 16xy + 4x + 8y + 16$$

Set the limits for the y-axis to $[-9, 9]$.

Create a function that makes the plot. Name the function **Prob6_A_5**.

Problem 6-A.6 (Section 6-1)

Use **ezplot** to plot the following implicit expression of two variables:

$$x^2 + y^2 = 1$$

Turn the grid on. Adjust the plot range so that your plot is easily readable. Add a title to your plot and label the axes.

Name the function that creates the plot **Prob6_A_6**.

Problem 6-A.7 (Section 6-1)

Graphically determine the value of $\cos(t)$ when $\cosh(t) = 3$.

Write a function, **Prob6_A_7**, that takes two inputs: **Lower** and **Upper**. These are the bounds on t used in the making the plot. Experiment with calls to your function until you can easily see the graphical solution to this problem. In your function, use **ginput** to capture the value of the graphical solution, and return the answer as an output variable from your function.

Problem 6-A.8 (Section 6-1)

Graphically determine the value of $f(x)$ given by:

$$f(x) = \sin(x)e^{-0.1x}$$

when

$$g(x) = -2x^2 + 5x + 4 = -15$$

Use the same general method you used for Problem 6-A.7. Name your function for this problem **Prob6_A_8**.

Be careful with this problem. You should return multiple answers, not a single answer.

Problem 6-A.9 (Section 6-1)

Use **ezplot3** to create a 3-D curve for the following relationship in the interval $[-9\pi, 9\pi]$:

$$x = e^{0.1t}\cos(t)$$
$$y = e^{0.1t}\sin(t)$$
$$z = t$$

Write a MATLAB function that creates the specified function. Name your function **Prob6_A_9**. In your function, call **rotate3d** so you can explore the plot from different viewing angles.

(You will need to look up **ezplot3** and **rotate3d** in MATLAB Help.)

Problem 6-A.10 (Section 6-2)

The data table below shows power dissipation for varying magnitudes of electric current in a circuit.

Current (amperes)	0	5	10	15	20	25
Power Dissipation (watts)	0	175	700	1575	2800	4375

Plot the data. Mark the data points with circles and connect them with a red solid line. Don't forget to label the x- and y-axes and to create a title for your graph.

Save the figure you generate as **Prob6_A_10.fig.**

Problem 6-A.11 (Section 6-2)

Steel is often tempered before it can be used for construction. Tempering relieves the internal stresses, reducing the brittleness while softening the steel. Tempering is carried out by heating the steel to a specific temperature, holding the temperature for a required length of time, and cooling the steel. The following table gives the temperature of a steel plate as measured during a tempering process.

Hours	0.0	0.5	1.0	1.5	2.0	2.5	3.0	3.5	4.0	4.5	5.0	5.5	6.0	6.5
°F	22	221	356	414	412	416	415	408	291	126	87	42	35	22

Plot the data by marking the data points with triangles and connecting them with magenta solid lines. Label the x- and y-axes and title your graph. Set the font size for axis labels and the title to 14 and 16, respectively.

Save the figure you generate as **Prob6_A_11.fig**.

Problem 6-A.12 (Section 6-2)

Hybrid vehicles combine more than one source of power to achieve higher power (e.g., locomotives) or longer duration of operation (e.g., submarines). Regenerative braking is one engineering method to enable a gasoline engine/electric battery hybrid vehicle to achieve higher traction at lower speeds and higher range for a given storage battery size. When a car is moving, it has an amount of energy dependent on how fast it is moving: the car's kinetic energy. When the car is stopped by the driver applying the brakes, the kinetic energy must be dissipated. What happens is that the energy of motion is converted into heat energy by the brakes, and the brakes become hot. In advanced automotive design, regenerative braking is used to store some of the energy of motion as the brakes are applied; not all the kinetic energy is wasted in heating the brakes, rather some is stored in an electrical battery.

An automotive company is in the process of designing a new gasoline engine/electric battery hybrid vehicle. From computer simulation using batteries with different storage capacities, data in the following table are found.

Bat. Capacity (ampere-hours)	10	20	30	40	50	60	70	80	90	100	110	120
Range (mi)	330	365	379	384	389	394	398	399	400	401	399	395

Plot vehicle range versus battery capacity. Determine an optimum value for battery capacity visually. Save your plot as **Prob6_A_12.fig**. On your graph, mark the optimum battery capacity using the **gtext** function.

What factors can you suggest that would make an optimum value, i.e., why isn't the relationship between battery capacity and range monotonically increasing? Put your answer into one or two concise paragraphs.

Problem 6-A.13 (Section 6-2)

The volume of a cone can be calculated using the following equation:

$$coneVolume = (\pi r^2)\left(\frac{coneHeight}{3}\right)$$

r is the radius of the base circle. Write a MATLAB function that inputs a vector of base radii and a scalar cone height and plots the cone volume as a function of base radius. Be sure you label the chart appropriately. Name your function **Prob6_A_13**.

Try your function with 0.1, 0.2,..., 10 feet for the radii. Make the cone height five feet.

Hint: If you are getting the error message 'Matrix dimensions must agree', go back to Chapter 5 and refresh your memory of cell-by-cell operations.

Problem 6-A.14 (Section 6-2)

Create a function, **Prob6_A_14**, that plots the following mathematical function in the range $a < x < b$ using **plot**:

$$f(x) = \frac{1}{xe^{-0.04/x}}$$

Test your function for the case a=**1**, b=**2**.

Problem 6-A.15 (Section 6-2)

Mathematical function $S(x)$ is defined as follows:

$$S(x) = 5\left(\frac{x}{0.2}\right)^{0.7}\left(\frac{1-x}{0.2}\right)^{-0.7}$$

Create a function, **Prob6_A_15** in the range $[a, b]$ using **plot**.

Problem 6-A.16 (Section 6-2)

A function $G(x, y)$ of two independent variables is defined as follows:

$$G(x, y) = \frac{x(7.14 - 0.26y)^{2.1}}{y^y}$$

Write a function, **Prob6_A_16**, that plots $G(x, y)$.

Since we have not studied plotting functions of two variables, you will have to figure out a path to show a number of plots that can give the sense of G

Refer to Figure 6-7. In it, you plotted a family of curves in one window. Use the methods you learned to produce Figure 6-7 to produce a family of G-plots for the following:

$$2 < x < 4$$
$$y = 2, y = 3, y = 4$$

Call the family of plots **Family_fixedY**. Refer to Figure 6-9. In it, you plotted several curves, each in its own window pane. Use the methods you learned to produce Figure 6-9 to produce a second window pane in the same figure and in it create a second family of G-plots this time for the following:

$$2 < y < 4$$
$$x = 2, x = 3, x = 4$$

Call the family of plots **Family_fixedX**. Your function **Prob6_A_16** should input the following arguments to generalize from the above situation. The inputs to **Prob6_A_16** should be the following:

> **lowerLimit** with the lower limit on x for **Family_fixedY** and the lower limit on y for **Family_fixedX**

> **upperLimit** with the upper limit on x for **Family_fixedY** and the upper limit on y for **Family_fixedX**

Your function **Prob6_A_16** should produce three curves in both families, **Family_fixedY** and **Family_fixedX**. The fixed values should be **lowerLimit**, **(lowerLimit+upperLimit)/2**, and **upperLimit**.

Problem 6-A.17 (Section 6-2)

Another function $G(x, y, z)$ of three independent variables is defined as:

$$G(x, y, z) = \frac{ze^{-0.3/x}}{\ln(y)\sqrt{x}}$$

Write a function, **Prob6_A_17**, that takes no input variables but that creates a three-pane vertical subgraph. The three subgraphs will be a plot of $G(x, y, z)$ subject to the following:

Pane A
$$0.1 < x < 4$$
$$y = 5, z = 3$$

Pane B
$$2 < y < 4$$
$$x = 0.8, z = 3$$

Pane C
$$0 < z < 4$$
$$x = 0.8, y = 5$$

Problem 6-A.18 (Section 6-3)

The following sample data represents the number of drivers that were ticketed on a given freeway in Michigan for traveling over 75 mph and for traveling under 55 mph.

Create two column vectors: **over75mph** and **under55mph** that each have twelve elements as given in the data table above.

Create a function **Prob6_A_18(over75mph, under55mph)**.

	Jan	Feb	Mar	Apr	May	Jun	Jul	Aug	Sep	Oct	Nov	Dec
Over 75 mph	38	29	43	51	67	84	79	95	73	55	46	19
Under 55 mph	12	11	9	3	4	2	0	1	3	7	11	14

Your function should generate an overlay plot of high and low speeders versus months (use month numbers). Mark the data points for **over75mph** with triangles pointing up and **under55mph** with triangles pointing down. Use different colors and widths of line for connecting the data points. Label your graph appropriately and generate a legend.

Problem 6-A.19 (Section 6-3)

Consider the following abstract function of temperature T (in degrees Celsius) as a function of time (in minutes:

$$T(t) = A \ln(t) - Be^{Ct}$$

Write a function **Prob6_A_19(T1_constants,T2_constants)**, whose input variables are each vectors describing a concrete version of $T(t)$. Suppose the following two inputs:

```
T1_constants = [A1, B1, C1];
T2_constants = [A2, B2, C2];
```

One input **T1_constants=[A1,B1,C1]** describes one concrete version of T, and the other input **T2_constants = [A2, B2, C2]** describes a second concrete version of T.

Your function **Prob6_A_19** inputs parameters for two distinctly different versions of $T(t)$, **T1** and **T2**.

Your function should graphically determine the intersection point(s) for **T1** and **T2**.

Write a short paragraph suggesting what physical situation $T(t)$ may capture.

Problem 6-A.20 (Section 6-3)

Consider the following function $F(x,y)$:

$$F(x, y) = \frac{x(7.14 - 0.26y)^{2.1}}{(y - x)^y}$$

Write a function to create a plot showing a family of $F(x,y)$ curves defined as follows:

$$2 < y < 6$$
$$x = 0.0, x = 0.5, x = 1.0, x = 1.5$$

How many curves should you overlay?

Use a line width of 4 for each of your curves. Be sure to include a legend.

Save your figure as **Prob6_A_20.fig**.

Problem 6-A.21 (Section 6-3)

Write a function, **Prob6_A_21**, to determine the intersection point graphically of the following curves:

$$F(x) = e^{-Bx}$$
$$G(x) = 1 - e^{-Bx}$$

Function **Prob6_A_21** should input **B** and should use **ginput** to capture the coordinates of the intersection point. Finally, **Prob6_A_21** should return one vector containing the **x** and **y** coordinates of the intersection point.

By running **Prob6_A_21** with values of **B** of **[0.5,1.0,2.0]**, understand how the intersection point varies with values of **B**. Write a short, concise paragraph describing your observations.

Problem 6-A.22 (Section 6-3)

The following sample data represents the number of drivers that were ticketed on a given freeway in Michigan for traveling over 75 mph and for traveling under 55 mph.

	Jan	Feb	Mar	Apr	May	Jun	Jul	Aug	Sep	Oct	Nov	Dec
Over 75 mph	38	29	43	51	67	84	79	95	73	55	46	19
Under 55 mph	12	11	9	3	4	2	0	1	3	7	11	14

Write a function to create a 1 by 2 subplot. Create bar plots of high and low speeders. Arrange the widths of bars so no gap is in between. Name your function **Prob6_A_22**.

Problem 6-A.23 (Section 6-3)

Engineers often use the *small angle approximation* to simplify their calculations. An example application of small angle approximation is the calculation of angular displacement for small motions of a pendulum.

Here is one version of the small angle approximation:

$$\sin(x) = x$$

This proves to be a reasonable approximation provided x is given in radians.

Write a function, **Prob6_A_23**, to demonstrate graphically the validity of this approximation for angles from 0 to 40 degrees.

In a single figure, show the three separate plots:

a) An overlay plot of $y1 = \sin(x)$ and $y2 = x$ in the interval $[0, 40]$ degrees. Put a legend on your plot.

b) A plot of the absolute error, i.e., $y = abs(\sin(x) - x)$

c) A plot of the relative error, i.e., $y = abs\left(\dfrac{\sin(x) - x}{\sin(x)}\right)$

Write a short, concise paragraph discussing the results you see in the graph generated by your function.

Problem 6-A.24 (Section 6-3)

Another version of the small angle approximation involves the cosine:

$$\cos(x) = 1 - \frac{x^2}{2}$$

Write a function, **Prob6_A_24**, to create analogous subplots to those required in Problem 6-A.23. This time, find the limit on the angle if the relative error must stay within one percent. In your relative error plot, draw a horizontal red dotted line at one percent error. Label the intersection of this line with the curve as the upper limit using **text** or **gtext**.

Problem 6-A.25 (Section 6-3)

A numerical approach to plotting functions (using **plot**) may not always produce accurate results.

Create an overlay plot of $y = \tan(x)$ in the interval $[-\pi, \pi]$ using **ezplot** and **plot**.

What is the value of $\tan(x)$ when the angle is $\pi/2$ or $-\pi/2$? Which curve reflects the behavior of $\tan(x)$ more accurately?

Write a concise paragraph explaining your observations.

Problem 6-A.26 (Section 6-4)

Use **ezplot3** to make a 3D line plot of the following system of equations in the closed interval $[0,5\pi]$

$$x = \cos(t)$$

$$y = \sin(t)$$

$$z = e^{-t}$$

Write a function that takes the range of **t** as a vector input and produces this plot. Name the function **Prob6_A_26**.

Problem 6-A.27 (Section 6-4)

Use the **meshgrid, griddata** and **mesh** functions of MATLAB to create a plot of the following dataset.

Table 6-3: Dataset of three variables

x	y	z
2.50	2.88	0.00
-0.38	-2.25	1.72
-2.62	-2.62	0.34
-2.13	2.13	-0.29
1.38	1.25	0.35
-2.38	1.38	-2.19
-2.13	2.88	3.25
2.63	3.00	-0.46
-0.38	-2.75	2.32
-1.88	-0.13	1.42

Set the marker color to green, and the marker size to 12. Set the font weight and size for all labels to bold and 14. Choose the 'copper' color map.

Problem 6-A.28 (Section 6-4)

Use **ezsurf** and **ezmesh** to make a 3D surface plot of the following function:

$$z(x, y) = 6(y^2 - x^2)^3 + (1 - y)^6$$

Write a function to solve this problem. Your function should take no inputs and produce two graphs, one for **ezsurf** and one for **ezmesh**. Name the function **Prob6_A_28**.

Problem 6-A.29 (Section 6-4)

Modify your solution for **Prob6_A_28**, and this time show both the surface and contour plots on the same graph. Set the transparency constant of the surface to 0.9. Set the shading to 'flat' and color map to 'cool'. Name your function **Prob6_A_29**.

Hint: Search for **surfc** in MATLAB help. Your graph should look similar to the one shown below.

Surface and Contour Plots for Problem 6-A.29

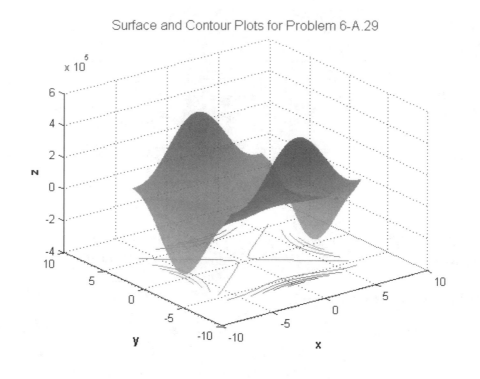

Problem 6-A.30 (Section 6-4)

Given **z** for the independent variables **x** and **y**:

$$z(x, y) = 5\left(x^3 + y^4 - x - \frac{y}{3}\right)e^{(-x^2 - y^2)}$$

Make a contour plot of **z**.

For this problem, write a function **Prob6_A_30** that takes two vectors as input: the interval for **x** and the interval for **y**. Your function should output a contour graph for the given range of **x** and **y**. Try your function with varying inputs to find the most meaningful range. Your result should resemble the graph given on the right.

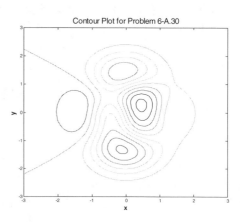

Contour Plot for Problem 6-A.30

Problem 6-A.31 (Section 6-4)

A helix in the 3D coordinates is defined as:

$$x = \cos(t)$$
$$y = \sin(t)$$
$$z = t$$

Write a function that takes no inputs and plots this helix in the range $[0,8\pi]$. Use **plot3** in your solution. Put a grid on your plot.

Name your function for this problem **Prob6_A_31**.

Problem 6-A.32 (Section 6-4)

Plot the following function using **surf**.

$$z(x, y) = x^3 - 5xy^2$$

Adjust the axes for **x** and **y** and make sure you show the characteristics of the surface. Write a function **Prob6_A_32** that takes the ranges for **x** and **y** as vectors and produces the plot.

When finished, your graph should look like the one below.

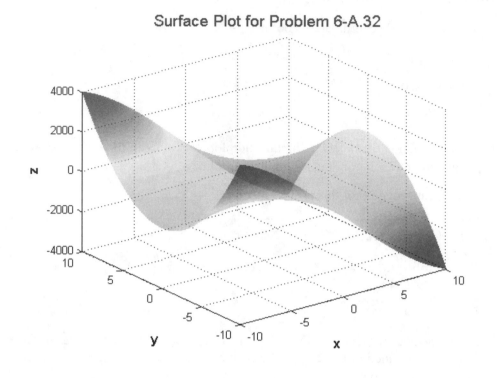

Surface Plot for Problem 6-A.32

Problem 6-A.33 (Section 6-4)

Plot a Mexican Hat Surface defined by the following function:

$$z(x, y) = \left(e^{-(x^2 + y^2)/0.3} - 0.4e^{-(x^2 + y^2)/7} \right)/0.6$$

Write a function called **Prob6_A_33** that takes no inputs and produces the plot. A Mexican Hat surface looks like the following:

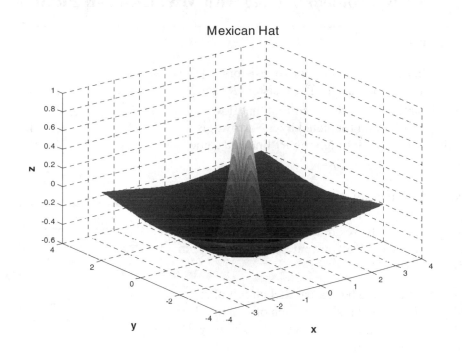

6-6.2   Set B: Problem Solving with MATLAB 2-D Plotting

Problem 6-B.1

A parallel circuit containing two resistors is shown to the upper right. The resistors R_1 and R_2 are in parallel because the current that flows through the circuit (depicted by the arrows) is split, some flows through the R_1 line and some goes through the R_2 line.

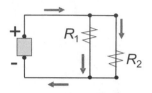

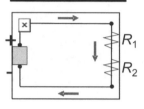

resistors in parallel

Parallel circuits are contrasted with series circuits. In a series circuit like the one shown to the lower right there are the same two resistors. In this case though all current flows through each of the resistors.

resistors in series

Suppose you have built a parallel circuit of n identical electrical resistors, each of r ohms. The effective resistance in the circuit R is given by the following equation:

$$R(r, n) = \frac{r}{n}$$

Plot equivalent resistance versus number of resistors for $R(r, n)|_{r = 100}$ in the interval for n $[1, 20]$. The symbol $R(r, n)|_{r = 100}$ means the function R with r fixed at 100 ohms.

Use an appropriate plot style (a phrase like '1.5 resistors' is meaningless). Fully label your graph.

Save your plot as **Prob6_B_1.fig**.

Problem 6-B.2

Coaxial cable is constructed by placing a core of conducting material inside an insulating material, then placing that inside a hollow cylindrical conductor. A cross-section for a coaxial cable is sketched to the right, where R is the outer radius of the insulating ring, r is the inner radius of insulating ring, and ε is the permittivity of the insulator, a physical constant dependent on the type of insulator. The permittivity is a measure of how well the material insulates.

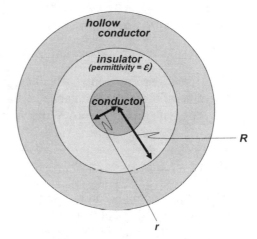

The capacitance per unit length C (farads/m) of a coaxial cable can be computed using the following relation:

$$C = \frac{2\pi\varepsilon}{\ln\left(\dfrac{R}{r}\right)}$$

Remember: the mathematical function $\ln$ is the natural logarithm.

Write a function, **Prob6_B_2_core**, that inputs ε, R, and a vector holding the lower and upper limit on values of r. Your function should examine the effect on C of changes in r for given values for ε and R by creating an appropriate plot.

Test your function by calling it with the following:

```
epsilon = 2.0364 x 10⁻¹¹;        % farads/m
R = 4;                           % mm
r_range =   [0.1, 3.9];          % mm
Prob6_B_2(epsilon, R,   r_range)
```

Write another function, **Prob6_B_2_top**, to examine the effect of both R and r on the unit capacitance of a coaxial cable. Do this by supplying **Prob6_B_2_top** test values to use for R; four test values will be supplied for **R**. **Prob6_B_2_top** will set up an overlay plot, one plot for each value of R will be put into the chart. **Prob6_B_2_top** should call **Prob6_B_2_core** to put on each one of the R-value plots. Test **Prob6_B_2_top** by calling it with the following:

```
epsilon = 2.0364 x 10⁻¹¹;                    % farads/m
R_testVal = [4.0, 4.15, 4.30, 4.50];        % mm
r_range =   [0.1, 3.9];                       % mm
Prob6_B_2_top(epsilon,R_testVal,r_range)
```

Do we need to worry about unit conversion from m to mm in this problem?

If you call your function with values such that **r > R** for some instances, what will the value of the unit capacitance be for those cases (greater or less than 0)? What is wrong with this question? Write your answers in a concise paragraph.

Problem 6-B.3

Implicit polynomials can be used for object recognition in images, for defect detection in quality control problems, and for interpretation of magnetic resonance images (MRIs). The following contour data are extracted from an image of a butterfly.

x	0.0	−4.0	−14.0	−23.0	−23.0	−19.0	−5.0	0.0	6.0	17.0	24.0	21.0	6.0
y	−2.0	−12.0	−20.0	−11.0	11.0	19.0	14.0	2.9	13.0	20.0	0.0	−18.0	−13.0

A pattern recognition specialist derives the following implicit polynomial that represents the above data:

$$f(x, y) = 0.000032044x^4 + 0.000016821y^4 - 0.0185x^2$$

Do an overlay plot of the data and the function $f(x,y) = 0$. Show the data points with blue pentagrams. How accurate is the implicit polynomial model in representing the butterfly data?

Label your plot appropriately.

Problem 6-B.4

MATLAB functions **tic** and **toc** are used for timing the execution of operations. The function **display** prints a string to the MATLAB Command Window. The function **input** prompts the user to supply a value. Look up these functions in MATLAB help.

Write a MATLAB function named **Prob6_B_4** that performs the following:

Part A	Displays a random integer number to the user. Utilizes **round(100*rand)** to generate the random number, and look up **round** and **rand** so you understand how they are being used.
Part B	Starts a stopwatch.
Part C	Asks the user to input *three times* the number displayed (from Part A).
Part D	Stops the stopwatch when the user enters an answer.
Part E	Displays the number of seconds the user took to enter the answer.
Part F	Checks if the user is correct. Displays 1 if the answer is right, 0 otherwise.

Problem 6-B.5

The height h (m) and horizontal distance x (m) traveled by a ball thrown at an angle α (degrees) with an initial speed v_0 (m/s) are given by the following relation:

$$x(t) = v_0 \cos(\alpha)t$$

$$h(t) = v_0 \sin(\alpha)t - \frac{1}{2}gt^2$$

where g is the acceleration due to gravity (9.81 m/s/s).

The ball will hit the ground at time t_{hit} (sec):

$$t_{hit} = 2\frac{v_0}{g}\sin(\alpha)$$

Part A   Create a function, **Prob6_B_5a**, that inputs the initial speed and angle of the ball's flight. Your function should plot height versus distance traveled by the object in the interval $[0, t_{hit}]$. Your function should return t_{hit}. Test your function by calling it with an initial speed $v_0 = 50$ m/s, and an initial angle $\alpha = 65°$.

Part B   Create a second function, **Prob6_B_5b**, whose functionality is like **Prob6_B_5a** except the plots show green if the height is less than or equal to 40 meters, cyan if the height is greater than 40 meters and less than or equal to 80 meters, and blue if the height is greater than 80 meters.

Problem 6-B.6

The height h (m) of the water in a leaking tank as a function of the time t (sec) at which the leak started can be modeled with the following relation:

$$h(t) = h_0 e^{-Bt}$$

Parameter B depends on the geometry of the tank and the area of the opening, and the constant h_0 is the initial ($t = 0$) height of the water. B can be a function of h, but for this problem, we will consider only constant values of B.

Explore the effect of B on $h(t)$ by following the directions below.

Write a function, **Prob6_B_6**, that creates an overlay family of plots of $h(t)$ for varying values of B ... **B=[0.1,0.2,0.4,0.8]**. Write your function in a general fashion so that it takes the following arguments:

    Prob6_B_6(h0, bVector)

bVector is a vector holding four values of B.

Describe in a concise paragraph your observation of the effect of differing constant values of B on $h(t)$.

B depends on the area of the hole producing the leak and on the geometry of the tank. The dependency on the hole area is easy to see: The larger the hole, the faster the leak. But see if you can determine why B would depend on the solid geometry shape of the tank.

Problem 6-B.7

The management team of the Edsel automotive company is planning to expand one of its plants by adding a new assembly line for sport utility vehicles (SUVs). The cost of setting up the new SUV assembly line is estimated to be $7 million.

The cost of manufacturing (raw materials, labor, etc.) an SUV is $36,000 and the company is planning to sell each SUV for $38,500.

Industrial engineers of the company say that the new assembly line will be capable of manufacturing nine SUV units a day.

How many days will it take for the company to be at the break-even point?

Create a function, **Prob6_B_7**, to solve the problem graphically. Plot a cost (fixed cost plus manufacturing cost) and a revenue curve. Label the break-even point using **gtext**.

Problem 6-B.8

Solve the previous problem assuming that the unit price for an SUV drops as the quantity manufactured increases. The fixed and manufacturing costs remain the same, but the production quantity Q and the selling price P are related by the following equation:

$$Q(P) = 9500 - 0.10P$$

Write a function, **Prob6_B_8**, that solves the problem. In addition to revenue and cost curves, plot as well the profit curve and label the point of maximum profit. Your function should return the number of SUVs your new line should create to produce maximum profit.

What do you suppose is the underlying reason for the relation between Q and P?

Problem 6-B.9

You are the manager of the GSXR motorcycle group for Black Hawk Motorcycle Company. Under your direction, a new model, the 1000, has been designed with 24 prototypes built. The following data represent the average speeds of the GSXR1000 prototypes around the Willow Springs race track:

Motorcycle	1	2	3	4	5	6	7	8	9	10	11	12
Speed (mph)	120	83	134	76	94	135	126	104	88	67	94	101
Motorcycle	13	14	15	16	17	18	19	20	21	22	23	24
Speed (mph)	134	98	93	113	168	92	134	149	83	99	106	170

Suppose as manager of the GSXR group, you have been charged with producing the 1000 model to have a specified speed capability on courses like Willow Springs.

Write a function, **Prob6_B_9**, that inputs your desired average speed at Willow Spring (in mph) and a vector containing the average speeds of the 24 prototypes. This function will produce a plot in which the speeds greater than or equal to the spec speed are plotted with blue squares, and others are plotted as red triangles.

Problem 6-B.10

A catenary is a uniform cable, fixed at its two ends, which supports its own weight. Examine the sketch to the right. A catenary cable (shown as solid) is attached at points P_1 and P_2. The attachment points are at a height h_1 and h_2 along an arbitrary y-axis. The xy-coordinates are arbitrary except that the y-axis must align with the down direction and the x-axis must have its zero point at an x-value corresponding to the lowest point of the catenary cable. The lowest point of the catenary cable is at a y-value of b.

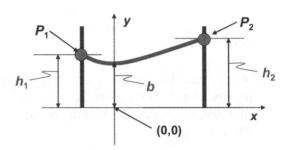

In the coordinate system described, the path of the catenary cable can be described as the following:

$$y(x) = b\cosh\left(\frac{x}{b}\right)$$

Write a function, **Prob6_B_10**, that inputs (a) the x-coordinate of P_1, (b) the x-coordinate of P_2, and (c) the value of b. Your function should return the heights of the two attachment points: the y-coordinate of P_1 and the y-coordinate of P_2. In addition, your function should create a plot of the catenary curve that the cable follows. Label the height of the two endpoints; use **text** for that labeling.

Test your function with the following:

```
P1_x = -30;    % m
P2_x = 50;     % m
b = 15;        % m
[P1_y, P2_y] = Prob6_B_10(-30, 50, 15)
```

The graph you generate should look similar to the one below.

Intuitively, when the two *x*-coordinate attachments points are equal in magnitude, what should be true about the *y*-coordinates of the two attachment points? Use your function to test your intuition.

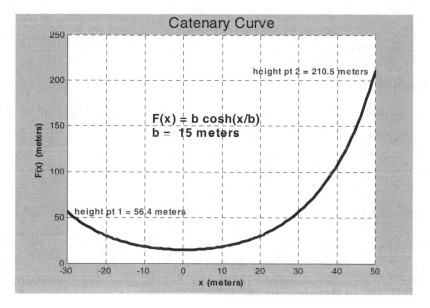

Problem 6-B.11

Run the MATLAB demo **graf3d** and experiment with various plot types, shadings and color maps. Make sure you try each plot type using the drop down menu.

Now run the teapot demo (**teapotdemo**) and observe the capabilities of MAT-LAB in producing graphics. Also practice with the options on the menu, such as the 'Orbit Camera', 'Orbit Scene Light' and 'Toggle Scene Light'.

Problem 6-B.12

Robots are used in various tasks that are difficult, unpleasant, or dangerous for humans. One such robot is built for the process of removal of rust on bridges, which produces chemicals dangerous for humans.

The rust-removing robot is programmed to follow a certain path on the trusses of the bridge. This path is defined as:

$$x = u$$

$$y = \sin(u)$$

$$z = \cos(u)$$

when $0 \leq t < 8\pi$ seconds, and

$$x = \sin(u - \pi)$$

$$y = u - 8\pi$$

$$z = \cos(u)$$

when $8\pi \leq t < 16\pi$ seconds.

Write a MATLAB function that takes no inputs and plots the path of the rust-removing robot for $0 \leq t < 16\pi$ seconds. Use overlay plots in your function to show the entire motion on a a single graph. Put a legend to distinguish between the two intervals of time.

Problem 6-B.13

A submarine crashed somewhere in the Pacific Ocean and a very secret government agency calls for your help. Geologists modeled the bottom of the ocean in the 8 x 8 sq mi area where the submarine is believed to have crashed. You are given this highly classified model:

$$z = 30e^{(-x^2 - y^2 - 2y - 1)} - 100(-x^3 - y^5)e^{(-x^2 - y^2)} - 985$$

where **x** and **y** are the coordinates in miles on the surface of the ocean, and **z** is the depth (feet) of the ocean at the coordinates. The bottom of the ocean looks like this in 3D coordinates:

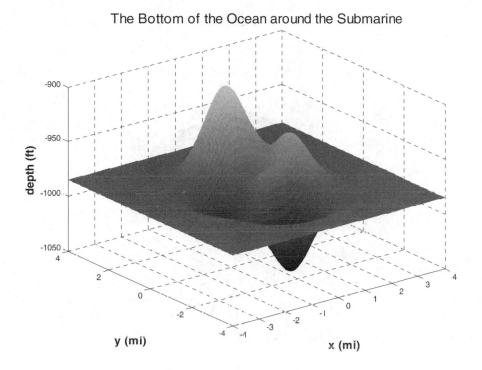

The Bottom of the Ocean around the Submarine

Radio communication with the submarine revealed that the submarine is exactly at the deepest point in this area (GPRS is not working). Your mission is to find out where exactly and at what depth the submarine lies in this area.

Write a function called **Prob6_B_13** that inputs the range vectors for **x** and **y** (both [-4,4] in this case) and produces a 3D surface plot. Visually determine the location of the submarine.

Problem 6-B.14

Use MATLAB's 3D graphing tools to plot a sphere in a cone. Your figure should look like the sample shown. To make this plot, open the editor and start your code with a **close all** command. Now experiment with the **cylinder** command to produce a cone. Next, switch the **hold** on your figure, and try different configurations of **sphere** or **ellipsoid** commands to place the ball into the cone. When you are done, switch the shading to flat. See the Help document for **shading** to do that.

When done, turn on the camera toolbar with the **cameratoolbar** command. Now you have more options to apply on your figure. Experiment with 'Orbit Camera' and 'Orbit Scene Light' switches. Click on Insert/Light to add a new light. Try different configurations for the place, type, and the color of the light.

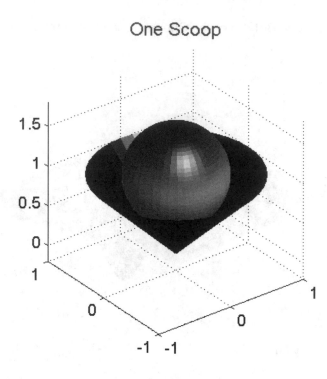

Problem 6-B.15

A plant scientist goes out to expedition in a remote island and discovers two new plant species. She takes measurements on the floral leaf length, floral leaf width, and petal length of the flowers of the plants. Now she needs your help to visualize the data and understand the nature of these new species.

She gives you the following sample to be analyzed:

Plant Type	Leaf Length (in)	Leaf Width (in)	Petal Length (in)
Species 1	4.2	0.6	1.2
Species 2	3.4	1.2	0.8
Species 2	3.9	1.2	1.5
Species 1	3.9	0.8	1.6
Species 2	3.9	1.3	1.2
Species 1	3.6	1.0	1.3
Species 1	4.3	0.7	1.3
Species 2	3.4	1.0	0.9

Write a script file to plot the leaf length, leaf width and petal length for the new plant species 1 and 2. Use different colored markers for data points for the two species of plants. Rotate the axis in 3D to visualize the data from varying angles.

Make sure your graph is appropriately labeled. Your graph should look like the one below when completed.

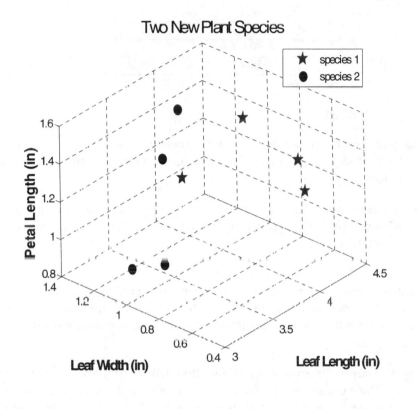

Problem 6-B.16

MATLAB offers you effective ways to discover computational models (functions) that represent experimental data. Why would it be useful to represent a dataset as a function? How could you use that function?

One tool for finding functions to "stand in for" a dataset is the *Fit Editor*.

Suppose you collected the data in the table below that includes (a) the stopping distance of a car and (b) the speed of the car.

Write a MATLAB function **Prob6_B_16** that takes no inputs and outputs the 7x1 vector of stopping distances. Your function should also plot the data using markers only (do not connect the data points). In the function, fully label your plot and put a grid on the figure.

Table 6-4: Dataset containing the speed of a car (independent variable) and the distance required to stop the car (dependent variable)

Speed (mi/h)	Stopping Distance (ft)
10.0	26.9
20.0	66.5
30.0	123.9
40.0	200.0
50.0	273.2
60.0	372.5
70.0	471.9

Run your function.

Once you have the figure, you can start experimenting with the *Fit Editor*. Click on Tools / Basic Fitting in your Figure window. This will bring up the Basic Fitting tool.

On the Basic Fitting window, find the 'Show Equations' checkbox and put a check in it. This will show the equation of any discovered function on your figure. Leave significant digits as 2.

Now select 'Plot Residuals'. Select the Bar Plot box, and switch it to 'Line Plot.' Residuals are the differences between your discovered function and the actual data points. The goal of *function discovery* is to both (a) minimize the residuals, and (b) to also to keep the residual distribution 'stable', i.e. not showing any trend across the data range.

You are ready to discover you first function that hopefully will represent the dataset. The more simple the function (the model for the data) the better (WHY?), so start with a linear function. Remember, linear functions are the functions of type **y=ax+b**. Look at the residuals. Do they seem to be stable? (Is there any trend you can see in the residuals?)

Now try a quadratic function. These are second degree polynomials. Select the quadratic polynomial and observe the residuals. Is the behavior of the residuals better now? (Is there less of a trend showing?)

Also try a cubic polynomial. You will observe that there is not much change in the trend of the residuals. But how about the complexity of the model you are using? A cubic is a third-degree polynomial, which does not substantially outperform a second-degree polynomial. Which model should we choose? (AGAIN - WHY is "simpler" to be preferred?)

At this point your figure should look like the following:

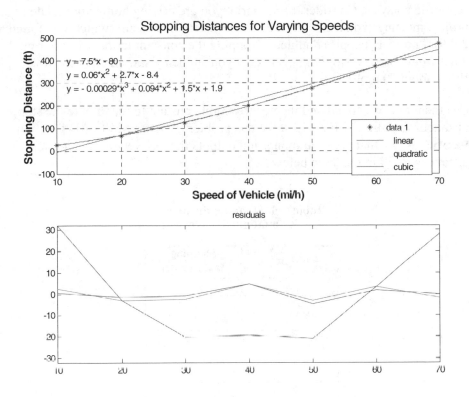

Problem 6-B.17

Reconsider the expression we found in **Problem 6-B.16** for the stopping distance of a vehicle. The best approximating function was:

$$y = 0.06x^2 + 2.7x - 8.4$$

In this equation, **x** is the speed in mi/h and **y** is the stopping distance in feet. The actual physical relationship for stopping distance of a speeding vehicle is

$$y = \frac{v^2}{2a}$$

where **a** is the deceleration of the vehicle. So, where are the other terms coming from? The constant term, -8.4 is very small compared to the stopping distances we measured so can be ignored as the measurement error, but the **2.7x** term can be as high as 189 ft when the speed **x** is 70 m/h. Can you think of a reason we may have such a 'delaying' effect on the deceleration of the vehicle?

If you said 'human reaction time', you were right. Now us use your findings for scientific analysis. Consider the scenario, you are driving home after a late night project meeting. You are so tired that your reaction constant (which you found as 2.7 for "normal" people) doubled. Dropping the constant term -8.4 in our model, what would be your stopping distance if you suddenly saw an obstacle at 10 mph? What about 70 mph?

Use the model we found and the new reaction constant to recompute the stopping distances for a tired driver. To solve this problem, write a MATLAB function **Prob6_B_17** that takes no inputs and outputs the 7x1 vector of stopping distances to fill in the table below.

**Table 6-5: Stopping distances
for a fatigued driver**

Speed (mi/h)	Stopping Distance (ft)
10.0	
20.0	
30.0	
40.0	
50.0	
60.0	
70.0	

Problem 6-B.18

Another very useful plotting tool in MATLAB is the Figure Palette. Before you start using this tool, run the functions you wrote for problems 6-B.16 and 6-B.17 to retrieve the stopping distances for alert and tired drivers in vectors named **stopAlert** and **stopTired**.

Now start Figure Palette by typing

```
>> figurepalette
```

This will bring up a fully interactive graphing tool!

Create a new subplot by clicking on '2D Axes'. Now, right click on the axes and select 'Add Data'. Choose the 'Plot Type' as scatter. Specify the X Data Source as [10:10:70], and Y Data Source as **stopAlert**. Redo this operation to add the **stopTired** data in an overlay plot.

Right click again on the axes and click on Properties. Put labels on the x and y axes and add a title to your figure. Use the Annotations pane to emphasize the difference between alert and tired drivers. Explore the tools to find how to put a legend on the figure.

When you are done, your figure should look the one below:

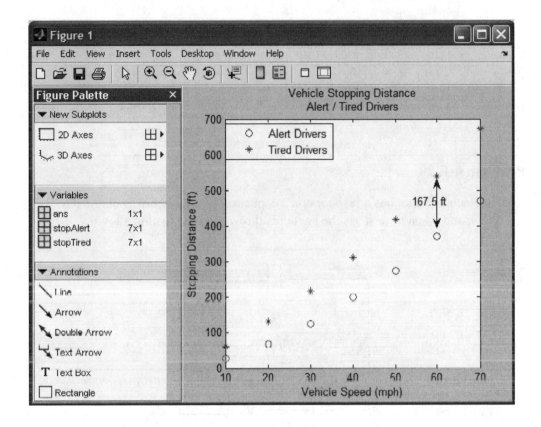

Problem 6-B.19

Anatolian Van Cats are a very unusual type of feline: they love water, and even are fond of swimming. The World Wildlife Association is attempting to find out if there is a danger that the Anatolian Van Cats are becoming extinct. The Association supports field work to conduct a census each year of these strange and wonderful cats in Anatolia. (If you don't know where Anatolia is ... look it up using Google.)

Plot the data in the table below using MATLAB and find a reasonable fit to the data using the Fit Editor. Using the mathematical fit you found, predict the populations in the year 1800, 1845 and 1910. Are your results realistic?

If you are getting results that seem totally out of range, increase the significant digits shown on the Fit Editor and try again. One point that is important to learn when using a relationship that is from fitting a function to a dataset is that you must be cautious in using the relationship for making predictions. It is essential when you use a function fit that you take every step you can to test your predictions for "reasonableness" and that you certainly do **not** accept results of such predictions blindly.

Table 6-6: Census counts for Anatolian Cats

Year	1810	1820	1830	1840	1850	1860	1870	1880	1890	1900
Count	22000	20000	19000	17000	15000	14000	12000	12000	11000	9000
Year	1910	1920	1930	1940	1950	1960	1970	1980	1990	2000
Count	9000	8000	7000	7500	6000	6500	5000	4500	4500	4400

Problem 6-B.20

An electric circuit has a resistor and a capacitor. The capacitor is charged to 0.2 Volts, and when time starts the switch is thrown to the position shown.

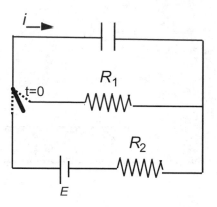

The electric charge at the capacitor starts to dissipate in time. The voltage at the capacitor is measured at fixed time intervals and the following data is obtained:

Table 6-7: Discharge of a capacitor as function of time for a given circuit

time (msec)	10	20	30	40	50	60	70	80	90	100
Voltage (V)	1.988	1.247	0.763	0.413	0.254	0.142	0.083	0.057	0.042	0.039

Plot the data using data markers and put a grid on your figure. Fully label the axes and put a title to your plot. Find a mathematical relation that represents the behavior of the voltage drop.

Food for further thought!

> If you are wondering how to discover a representation for the data with an exponential function, consider the following. Exponential functions are in the form $y = be^{mx}$. Taking the logarithm of both sides, we find $\log(y) = \log(b) + mx$. This implies, if we plot x vs. the logarithm of the data, we will obtain a linear behavior. Thus, we can discover a linear function and take the exponent of the constant term to discover the parameters of the exponential function.

Chapter 7

Arrays

Where are we on the road to mastering MATLAB? So far, you have learned how to do the following:

- Use MATLAB for scalar computations (Chapter 3)
- Save your work in MATLAB user-defined functions (Chapter 4)
- Debug MATLAB programs (functions) (Chapter 4)
- Use MATLAB for vector operations (Chapter 5)
- Use MATLAB to make 2-D plots (Chapter 6)
- Use the MATLAB Help facility to let you extend what you know about MATLAB (Chapter 6)

This progression has a pattern. You first learned to use MATLAB for problems that you are familiar with: scalar operations. You added to what you had learned about scalars and the MATLAB environment by creating scalar functions, and you saw how to test and debug your functions. The next step involved groups of scalars: vector operations. Going further, you applied your understanding of vectors to create two-dimensional plots. Finally, you saw how to use Help to refresh your memory on fine points (e.g., looking up the arguments needed by a built-in function) and to enhance your MATLAB capability (e.g., finding a MATLAB built-in function that will perform an operation you need).

The story line we have followed has been one of building on what you know. In this chapter, the progression is continued by introducing the topics of array and matrix operations in MATLAB. You have seen that a good way to think about MATLAB vectors is that they are ordered, linear groupings of scalars. Conceptually, arrays in MATLAB are a simple extension. Instead of having a one-dimensional grouping of scalars as in vectors, MATLAB arrays are two-dimensional groups of scalars. To access and operate on an element in a vector, you need one index, a number that tells you the location of the element in the vector. For example, to set the fifth element of a vector **A** to the number 100, you use the command **A(5)=100**.

Think of a MATLAB array as a table as in Excel.[1] Any given scalar number in such a table has two specifications that give the element's location in the table. In Excel, the two specifiers are a letter to indicate the column in which the scalar resides and a number to specify the row in which the scalar resides. For example, the location in a standard Excel table for the item in the third column, second row of the table is specified by **C2**, where the **C** indicates third column and the **2** indicates second row.

In MATLAB, to specify where a scalar is located in a table, two numbers are used. If **B** is a 2 x 4 table (read 2 by 4)—meaning that **B** has two rows and four columns—then use **B(2,3)** to access the scalar in the second row, third column. The first index number (**2**) is the row identifier; the second (**3**) is the column identifier.

You must link what you learned earlier about vectors, and what you are going to learn about arrays. In fact, vectors in MATLAB are special cases of arrays. A row vector can be thought of as a table with one row; a column vector can be thought of as a table with one column. What you have learned for vectors carries over to arrays with minimal addition. Remember that so far for vectors we covered only cell-by-cell multiplication, division, and exponentiation: **.***, **./**, and **.^**.

We will follow the same general outline as in Chapter 3 (scalars) and Chapter 5 (vectors) in this chapter but with a twist. That twist is that we will consider two types of operations: the cell-by-cell operations you are familiar with, and what we call matrix operations. Let's get some terminology straight first.

- The MATLAB data object that is analogous to an Excel table (two-dimensional table in this book) will be called an array.
- Two special cases of arrays as data objects have already been dealt with—row vectors and column vectors.
- For multiplication, division (right and left), and exponentiation, there are two types of operations that can be applied to MATLAB arrays: cell-by-cell operations and matrix operations. That is, there are two types of multiplication, two types of right division and left division, and two types of exponentiation.
- For all other operations on MATLAB arrays, there is only a cell-by-cell version.

The reason it is so important to get these terms and the concepts behind them straight is that the term *cell-by-cell operation* has another name, one that is used often in MATLAB Help. That other name is *array operation*. For example, if you look up the Help document for *Help/MATLAB/Functions-Alphabetical List/Arithmetic Operators*

---

1. In this chapter, only two-dimensional tables will be discussed. MATLAB supports three-, four-, ... *N*-dimensional tables.

you will find that cell-by-cell multiply (**.***) is called *array multiply* in Help. The use of the term *array multiply* (and *array right divide*, *array left divide*, *array exponentiate*) leads to a confusion between what term to use when referring to the data object versus the operation *applied to* the data object. To get around this problem, in this book we refer to the data object (the table) as an *array* and the two types of multiply operations which can be applied to the data object as *cell-by-cell multiply* and *matrix multiply*. Just keep in mind that MATLAB Help uses the term *array multiply* to mean an **operation** that in this book we call *cell-by-cell multiply*.

Below we will go over array creation, array access, and array transposing. All of these subjects are relatively straightforward extensions from your current knowledge of vectors. When we go on to MATLAB array built-in functions, operators, and expressions, we will make a clean distinction between cell-by-cell operations and matrix operations.

7-1   Array Creation

Start MATLAB and perform the following operation in the Command Window:

```
>> A = [1, 2, 3; 10, 20, 30]
```

You will see the following result:

```
A =
          1.00            2.00            3.00
         10.00           20.00           30.00
```

This is an application of the same square bracket operator that you know how to use to create vectors. But above, we have used a single facet that is new and that allows the creation of an array: a *semicolon*. A semicolon as punctuation in the square bracket operator tells MATLAB to start a new row.

Think of the single command `A = [1, 2, 3; 10, 20, 30]` as made up of two parts:

```
A1 = [1, 2, 3]
```

and

```
A2 = [10, 20, 30]
```

These are familiar operations with the square bracket operator to create row vectors. The operation to create **A** above was similar except there are two rows built, and the two rows are stacked on each other with the semicolon punctuation. And the stacking produces a 2 x 3 array. Using **A1** and **A2** above, the same **A** could be produced by the following:

```
>> A = [A1; A2]
```

When you first learned how to create vectors in MATLAB (Section 5-1), you saw there were three general methods used: square bracket operator, **linspace**, and the colon operator.[2] The square bracket operator is the only general method for creating arrays in MATLAB. The colon operator and **linspace** are used *only* to create vectors, not to create arrays directly. However, both can be used to create vectors that are subsequently composed into an array:

```
>> A = [1:3:15; linspace(0,1,5)]
A =
          1.00      4.00      7.00     10.00     13.00
             0      0.25      0.50      0.75      1.00
```

---

2. These three methods were introduced in Section 5-1. There are several others in MATLAB as well. See, for example, **logspace** in MATLAB Help.

Suppose you were asked to perform a variant on the array creation problem above: to create an array which had two columns, where the first column had elements that were **1:3:5** and the second column had elements that were **linspace(0,1,5)**. In this example, you are again asked to stack two vectors, but this time the vectors are column vectors and the stacking is to be left to right across the array. The following command will accomplish this task:

```
>> A = [(1:3:15)', linspace(0,1,5)']
A =
              1.00                 0
              4.00              0.25
              7.00              0.50
             10.00              0.75
             13.00              1.00
```

(1:3:5) The symbol **'** creates a column vector; note the use of the transpose operator that you learned in Section 5-3. Similarly, **linspace(0,1,5)'** creates a column vector. Most important is that stacking these two row vectors across, left to right, is accomplished with a comma punctuation. This is analogous to building a row vector.

One more mutation of this array creation problem can serve as a test of your understanding. Suppose you were asked to take the two column vectors, **(1:3:5)'** and **linspace(0,1,5)'**, and stack them on top of each other—that is, stack them in the up-down direction:

```
>> >> A = [(1:3:15)'; linspace(0,1,5)']
A =
              1.00
              4.00
              7.00
             10.00
             13.00
                 0
              0.25
              0.50
              0.75
              1.00
```

The semicolon is again telling MATLAB to create a new row but in this case after the new row is created, what is inserted is an entire column vector.

When you use the square bracket operator to create an array in MATLAB, you must be careful that the pieces you are assembling have consistent dimensions. For example, suppose you have two vectors **B1** and **B2**:

```
>> B1 = [1, 2, 3];
>> B2 = [10, 11];
```

When you stack them, you will see the following:

```
>> stackedUpDown = [B1; B2]
??? Error using ==> vertcat
All rows in the bracketed expression must have the
same number of columns.
```

Stacking in computer jargon is called *concatenation*. The operation that you wanted to apply with the semicolon punctuation inside the square bracket operator above is *vertical concatenation*. The MATLAB syntax error message is telling you that vertical concatenation cannot work in the case of **[B1; B2]** because the dimensions are wrong; **B1** is a 1x3 row vector and **B2** is a 1x2 row vector. They cannot be vertically concatenated because the resulting object would not be a valid array. Arrays must have the same number of elements in all rows of the array, and the same number of elements in all columns of the array.

As a final topic in this section, two special purpose built-in functions for array creation deserve attention. In the section on creating vectors (Section 5-1) you saw these two special-purpose vector-creation functions: one for creating a vector of all zeros and a second for creating a vector of all ones. At the time, the syntax of these two built-in functions may have puzzled you. For example, to create a row vector of six elements, all having the value 0 you used the following:

```
>> sixZeros_Row = zeros(1,6);
```

To create a column vector of six 0s, you used the following:

```
>> sixZeros_Col = zeros(6,1);
```

As you extend to arrays, the syntax for **zeros** and **ones** will make more sense. Both of these built-in functions create arrays. The first argument is the number of rows to be created, and the second is the number of columns to be created:

```
>> twoByFourZeros = zeros(2,4)
twoByFourZeros =
            0           0           0           0
            0           0           0           0
```

1. Semicolon punctuation inside the square bracket operator indicates to MATLAB that a new row is to be created.

2. When using the square bracket operator to create arrays in MATLAB, dimensions of the array pieces being assembled are consistent.

3. `ones` and `zeros` are built-in functions in MATLAB that create arrays whose elements are all value 1 or all value 0, respectively. Both `ones` and `zeros` take two arguments: the first is the number of rows in the array that will be created, and the second is the number of columns in the array that will be created.

Synopsis for Section 7-1

7-2   Accessing Array Elements

In Section 5-2, you learned how to access elements in vectors, fetching values of elements and setting values of elements. This section is a short extension to learn how the same access operations are performed on arrays. Access to an entire array is like access to an entire vector. For example, to set an array **A** to `[1,2;3,4]`, you must do the following:

```
>> A = [1,2; 3,4]
A =
            1.00              2.00
            3.00              4.00
```

Access to individual elements in an array is close to what you learned for vectors. The only conceptual extension is that arrays have two indices instead of the one index needed for vectors.

7-2.1   Fetching Elements of Arrays

Suppose you have an array **A** built by the following:

```
>> A = [1,2,3; 10,11,12]
A =
           1.00               2.00               3.00
          10.00              11.00              12.00
```

You want to pull out the value of the element at the second row, third column, and put that value in a scalar variable x. The following code accomplishes that task:

```
>> x = A(2,3)
x =
          12.00
```

Or suppose you want to set a vector variable **V** to the second and third elements in the second row of **A**:

```
>> V = A(2, [2,3])
V =
          11.00             12.00
```

Similarly, you could pull out the first and second elements in the second column of **A**:

```
>> Y = A([1,2], 2)
Y =
           2.00
          11.00
```

And as a final example for this first set, you could pull out the first and third elements of the second row of **A** by the following:

```
>> Z = A(2, [1,3])
Z =
        10.00            12.00
```

There is one MATLAB shorthand that makes it much simpler to extract elements from an array. Suppose you start with the array **B**:

```
>> B = [1,2,3,4; 10,11,12,13; 20, 21,22,23]
B =
        1.00        2.00        3.00        4.00
       10.00       11.00       12.00       13.00
       20.00       21.00       22.00       23.00
```

You want to extract the entire second column. This is how you can accomplish that task:

```
>> X2 = B(:, 2)
X2 =
        2.00
       11.00
       21.00
```

The colon used as an access element says *all* to MATLAB. In this case, since the colon appeared in the row index location, the meaning was *all rows*. Another use of the colon access would be to pull out an entire row of **B** as follows:

```
>> X3 = B(3, :)
X3 =
       20.00       21.00       22.00       23.00
```

When we fetch multiple elements from an array, the *shape* of the extracted object is retained: If we fetch an entire row, we get a row vector; if we fetch an entire column, we get a column vector.

Fetching elements from an array can result in another array. For example, suppose you use the same array **B** from above, and you want to fetch the entire third and fourth columns:

```
>> partOfB = B(:, [3,4])
partOfB =
        3.00            4.00
       12.00           13.00
       22.00           23.00
```

Or suppose you want to fetch the first and second elements in the second and third columns of **B**:

```
>> anotherPartOfB = B([1,2], [2,3])
anotherPartOfB =
          2.00              3.00
         11.00             12.00
```

So, array fetching is like fetching elements from vectors. The only difference is having two index elements for arrays when only one was needed for vectors.

7-2.2   Setting Elements of Arrays

Just as fetching elements of arrays is a relatively straightforward extension from fetching elements of vectors, so too setting elements of arrays is analogous to setting elements of vectors. Suppose again you have array **B**:

```
>> B = [1,2,3,4; 10,11,12,13; 20, 21,22,23]
B =
          1.00      2.00      3.00      4.00
         10.00     11.00     12.00     13.00
         20.00     21.00     22.00     23.00
```

You want to set the element at row 2 and column 4 to 100:

```
>> B(2,4) = 100;
B =
          1.00      2.00      3.00      4.00
         10.00     11.00     12.00    100.00
         20.00     21.00     22.00     23.00
```

Or suppose we wanted to create a new array **C** that is identical to **B** as just modified, except that the second and third columns are interchanged. This reversal of column 2 and column 3 is accomplished below:

```
>> C = B;
>> C(:,[2,3]) = B(:, [3,2])
C =
          1.00      3.00      2.00      4.00
         10.00     12.00     11.00    100.00
         20.00     22.00     21.00     23.00
```

Setting elements in arrays, including setting multiple elements with one MATLAB command, is easy—with one caution. The shape of the array to be set (indicated on the left-hand side of the assignment operation) must be the same as the shape of the array that holds the new values (indicated on the right-hand side of the assignment operation). In the example above, the target of the operation to set some elements of **C** was a three-row by three-column array: three rows because all rows of **C** are indexed, and two columns because we indexed only column two and column three. This matches the shape of the array on the right-hand side of the assignment operator, again a three-row by two-column array.

7-2.3   Use of *end* for Array Access and the *size* Built-in Function

Two final points are necessary for the discussion of array access. First, for vector access, the word **end** used as an index item has the meaning of *the last element in the vector*. The generalization of **end** holds for array access. Suppose someone handed you an array **D** but did not tell you the dimensions of the array (the number of rows and number of columns). Further, suppose you are asked to fetch the element at the last row, last column. The operation **D(end,end)** will accomplish that task.

Suppose that you are asked to set the elements in the next to last and last row, next to last column and last column to the 2x2 array **[100, 101; 200, 201]**. That can be accomplished by **D(end-1:end,end-1:end) = [100, 101; 200, 201]** even though you do not know the size of **D**.

Here is an example:

```
>> D = [1,2,3,4; 10,11,12,13; 20, 21,22,23]
D =

            1.00        2.00      3.00       4.00
           10.00       11.00     12.00      13.00
           20.00       21.00     22.00      23.00
```

Change the last and next to last row/column elements:

```
>> D(end-1:end,end-1:end)  =   [100, 101; 200, 201]
D =
            1.00        2.00      3.00       4.00
           10.00       11.00    100.00     101.00
           20.00       21.00    200.00     201.00
```

The index into **D** on the right-hand side of the assignment operation above used the vector creation colon. Do not confuse this with the colon used alone in an index meaning *all*.

Finally, one useful built-in function, often needed to find the dimensionality of an array is **size**. For vectors, we were able to find the number of elements in a vector using **length**. Recall the following example:

```
>> V = [1,2,3,99];
>> lenV = length(V)
lenV =
            4.00
```

This gives the number of elements in **V**. More generally though, for an array **A**, **length(A)** returns the length of the longest dimension of **A**. That is, if the number of rows is greater than the number of columns, then **length** returns the number of rows. But if the number of columns is greater than the number of rows, then **length** returns the number of columns.

Many times when working with an *MxN* array, knowing the values for *M* and *N* is useful. To find the number of rows and columns of an array **A**, the **size** built-in function may be used:

```
>> D = [1,2,3,4; 10,11,12,13; 20, 21,22,23]
>> [numRows, numCols] = size(D)
numRows =
            3.00
numCols =
            4.00
```

Note that **size** returns two output variables. To capture both, the calling syntax must include both as above.

1.   Array access operations (**fetch** and **set**) are directly analogous to vector access operations.

2.   For array setting, the part of an array to be set (right-hand side of assignment) and the elements which will be inserted (left-hand side of assignment) must be the same shape.

3.   The colon may be used as an index element to indicate *all*.

4.   **end** is used in array access as it is used in vector access.

5.   To determine the number of rows and columns in an array, use **size**.

Synopsis for Section 7-2

7-3   Transpose Applied to Arrays

In the case of vectors, the *transpose* operator was useful for converting row to column vectors or column to row vectors (Section 5-3). For the more general case of arrays, with row vectors and column vectors as special cases of NxM arrays, the transpose operator is used to flip an array. More formally, if $\mathbf{A}$ is an NxM vector, then $\mathbf{A}'$ will be an MxN array whose elements are defined by `A'(i,j) = A(j,i)`. The effect of applying the transpose operator to an array is to flip rows and columns: What was a row is now a column, and what was a column is now a row:

```
>> D = [1,2,3,4; 10,11,12,13; 20, 21,22,23]
D =
            1.00      2.00      3.00      4.00
           10.00     11.00     12.00     13.00
           20.00     21.00     22.00     23.00

>> transposeD = D'
transposeD =
            1.00     10.00     20.00
            2.00     11.00     21.00
            3.00     12.00     22.00
            4.00     13.00     23.00
```

---

1. The transpose operator (`'`) applied to array **A** has the effect of flipping **A** about its major diagonal: What was a row becomes a vector after the transpose operator is applied, and what was a column becomes a row.

Synopsis for Section 7-3

7-4   Array Built-in Functions, Operators, and Expressions

As we did for scalars in Chapter 3 and for vectors in Chapter 5, now you have learned for arrays: how to fetch arrays and elements in arrays, and how to set arrays and elements in arrays. The next step is to learn how to build commands using arrays. We will follow the same basic path for arrays that we followed before when you learned about scalars and vectors.

7-4.1   Array Built-in Functions

The discussion in Section 5-4.1 (Vector Built-in Functions) carries over just about wholesale. Two general kinds of array built-in functions exist: those that work in a cell-by-cell fashion on the input array and those that work on all elements of an array at once. This same point was made for vector built-in functions.

As an example of an array built-in function that works cell-by-cell, consider the hyperbolic **sin** function. Suppose we had an array of angles **A** (in radians):

```
>> A = [0:pi/8:pi/2; pi/2:pi/8:pi]
A =
            0      0.39      0.79      1.18      1.57
         1.57      1.96      2.36      2.75      3.14
```

Now suppose we apply **sinh** to **A**:

```
>> hyperbolicSin_A = sinh(A)
hyperbolicSin_A =
            0      0.40      0.87      1.47      2.30
         2.30      3.49      5.23      7.78     11.55
```

Array built-in functions that are cell-by-cell apply the corresponding scalar built-in function to each cell of the input array. If the target array is NxM, then the output will also be NxM.

The situation for array built-in functions that operate on all array elements at once is analogous to vector operators of the same type. But for array application of operators like those of Table 5-1, one point requires discussion.

To make the explanation concrete, let us consider the **sum** built-in function. Applied to a row vector or a column vector, **sum** sums the values of the vector's elements and returns a scalar value. But applied to an array, **sum** will output a vector, for example, with array **D**:

```
>> D = [1,10; 100,110]
D =
                1.00              10.00
              100.00             110.00
```

When you apply **sum**, you find the following:

```
>> sumD = sum(D)
sumD =
              101.00             120.00
```

The returned vector contains the sums that are over the columns of **D**. Summing the columns is the default operation for the built-in array function **sum**, and **sum** takes two arguments, the second one optional. The first argument is always the array to be summed over. The second argument indicates which dimension the sum is to be performed over. If the second argument is equal to 1, then the sum is carried out over columns, and a row vector is returned. If the second argument is equal to 2, then the sum is carried out over rows, and a column vector is returned. In the case of **D** given above, the following occurs:

```
>> sumOverColumns = sum(D,1)
sumOverColumns =
              101.00             120.00

>> sumOverRows = sum(D,2)
sumOverRows =
               11.00
              210.00
```

If no second argument is given to **sum**, then the second argument is assumed to be 1, e.g., the default is 1.[3]

The array built-in function **mean** works similarly to **sum**. That is, **mean** takes an optional second argument that specifies whether the mean is to be taken row-wise or column-wise. Not all array built-in functions work in this manner, however. MATLAB Help should be consulted for the arguments needed by array built-in functions when you are unsure.

---

3. As noted earlier (Footnote 1 on page 7-276) MATLAB supports N-dimensional arrays. Array built-in functions like **mean** take a second argument which indicates the dimension over which the operation is to be performed. Though we will not deal with arrays of dimension greater than 2, you can learn about higher dimensionality arrays in MATLAB Help.

7-4.2   Array Operators

The same four classes of operators exist (as described in the chapter on vectors) when we extend the discussion to arrays. These four classes are as follows:

1. **Special operators:** includes the assignment operator, the transpose operator, the colon operator, and the square brackets operator

2. **Arithmetic operators:** enables normal operations of arithmetic for two scalars, two arrays, or one scalar and one array

3. **Relational operators:** enables comparing two scalars, two arrays, or one scalar and one array

4. **Logic operators:** enables combining the results of one or more relational tests

This is the same basic set of four operator classes as you know for vectors.

The individual operators in Classes 1, 3, and 4 above extend directly to cover arrays. This is not too surprising since vectors are a special case of arrays. But Class 2 operations need to be extended. In the chapter on MATLAB vectors (Chapter 5), we limited discussion to cell-by-cell operations. In addition to cell-by-cell arithmetic operations, another possibility is matrix operations.

Extension to arrays for all but the arithmetic operators is straightforward. Remember, everything you learned about the relational and logic operations on vectors carries over to arrays. The point to keep in mind is that all relational and logic operations applied to arrays are cell-to-cell operations.

In the rest of this section, we discuss only the arithmetic operators (Class 2). The entire set of arithmetic operators for scalars, vectors, and arrays is shown in Table 7-1. This table is reorganized and expanded from the similar tables in the chapter on scalars (Table 3-3) and the chapter on vectors (Table 5-2). The reorganization is in two groupings. cell-by cell and matrix operations. We will cover these in turn below.

7-4.2.1   Cell-by-Cell Operations on Arrays

With one exception, you have seen all of the cell-by-cell operations earlier in Chapter 5 on vector operations. For example, apply the cell by cell multipli cation operator to two arrays **A** and **B**:

Table 7-1: Arithmetic Operator Class for Arrays

Type	Operator name	Operator Symbol
cell-by-cell	unary plus	**+**
cell-by-cell	unary minus	**–**
cell-by-cell	addition	**+**
cell-by-cell	subtraction	**–**
cell-by-cell	multiplication	**. ***
cell-by-cell	right division	**. /**
cell-by-cell	left division	**. **
cell-by-cell	exponentiation	**. ^**
matrix	multiplication	*****
matrix	right division	**/**
matrix	left division	****
matrix	exponentiation	**^**

```
>> A = [2:4; 20:10:40]
A =
          2.00          3.00          4.00
         20.00         30.00         40.00
>> B = [1:3; 1:3]
B =
          1.00          2.00          3.00
          1.00          2.00          3.00
```

This yields the following:

```
>> A .* B
ans =
```

2.00	6.00	12.00
20.00	60.00	120.00

The cell-by-cell exponentiate operator applied to obtain **A** raised to the **B** power cell-by-cell yields this:

```
>> A .^ B
ans =
```

2.00	9.00	64.00
20.00	900.00	64000.00

Finally, the right divide cell-by-cell operator applied to obtain **A** divided (right divide) by **B** yields this:

```
>> A ./ B
ans =
```

2.00	1.50	1.33
20.00	15.00	13.33

One item in the cell-by-cell entries in Table 7-1 is new: cell-by-cell left division. In the earlier chapters on scalars and vectors, you learned one form of the division operator. The correct name for the division operator you learned is right division. Right division is the standard numerical division operator that you have used through your math and science classes. **A ./ B**, means $\frac{A}{B}$. The cell-by-cell left division, **A .\ B** means $\frac{B}{A}$. Keeping left and right versions of the division operator straight is easy. The slash points to the right for Right Divide, and it points to the left for Left Divide. The item the slash points toward should be on the bottom of the fraction. Left cell-by-cell division is not often used by experienced MATLAB programmers, but you should remember it for completeness.

In general, the cell-by-cell array operations are all extensions of vector cell-by-cell operations that you already know. In the chapter on vectors, you learned two rules that must be satisfied for any cell-by-cell operation to be legally applied to two vectors **A** and **B**: The length of **A** and **B** must be the same, and **A** and **B** must be row or column vectors. Generalizing to arrays, the two rules collapse to one rule: The shape, or more technically, the dimensionality of **A** and **B** must be the same for any cell-by-cell operation to be legal. If **A** is an NxM array, then for any cell-by-cell operation to be legal between **A** and **B**, **B** must be an NxM array. So, the number of rows in **A** and **B** must be equal, and the number of columns in **A** and **B** must be equal. If you think about the meaning of cell-by-cell operation, you will see that this rule is a logical necessity.

We will end this section with an example that will help to solidify your thinking about cell-by-cell array operations. The example is drawn from an inventory/production type of accounting problem common to any engineering firm.

The ABC electronics factory makes four different items: a 48-inch HDTV, a 32-inch regular TV, a computer called the M2 model, and a DVD player called the R2 model. Table 7-2 lists three categories of cost for each product. Table 7-3 lists the production volume last year for each product ABC makes.

Table 7-2: ABC Production Costs for Each Unit Produced

Product	Material Cost ($)	Labor Cost ($)	Transportation Cost ($)
48-in HDTV	892	531	89
32-inch TV	149	128	15
M2 computer	175	75	15
R2 DVD	89	22	5

Table 7-3: ABC Production Volume Last Year

Product	Quarter 1	Quarter 2	Quarter 3	Quarter 4
48-inch HDTV	532	629	680	801
32-inch TV	431	329	387	522
M2 computer	1212	521	839	984
R2 DVD	534	545	498	577

Your goal is to compute (a) the total cost for materials used on all four product lines for each quarter and (b) the total yearly cost for materials used in each of four product lines.

As standard practice, your first task is to understand what the problem is asking you to do, and then to sketch out the path to solve the problem. Examine the two tables above. The material cost for ABC to make one 48-inch HDTV is $892 for each unit (from Table 7-2). In the first quarter, ABC manufactured 532 of these HDTV units (from Table 7-3). Part (a) of the problem statement asks you to compute the material cost total for all four product lines for each quarter. You know (from what we said above) that the contribution in Quarter 1 to material costs from the HDTV product line is 532 * $892 = $474,544. Think through the entire problem statement, and be confident you could solve the problem with just paper and pencil if you needed to.

You can see this problem is not conceptually difficult but is tedious because if you were to compute each contribution (as we just did for the Quarter 1 contribution to materials cost from the HDTV product line), you would need to carry out a number of scalar operations. MATLAB provides a better way.

As a first step toward a MATLAB solution, you need to represent the data in Table 7-2 and in Table 7-3. Creating arrays for the two ABC data tables is the most natural way to proceed:

```
>> unitCosts = [892, 531, 89; ...
                149, 128, 15; ...
                175,  75, 15; ...
                 89,  22,  5];
>> productionVolume = [532, 629, 680, 801; ...
                       431, 329, 387, 522; ...
                      1212, 521, 839, 984; ...
                       534, 545, 498, 577];
```

In **unitCosts**, each row contains data about an ABC product: Row 1 is the HDTV, Row 2 is the regular TV, etc. Each column in **unitCosts** is a cost category: Column 1 is material costs, Column 2 is labor costs, etc. In **productionVolume**, each row is again an ABC product, and each column is production data for one of the calendar quarters. **unitCosts** and **productionVolume** are lifted from Table 7-2 and Table 7-3.

In terms of the two MATLAB arrays defined above, here is the material cost for Quarter 1 for the HDTV line:

```
>> q1_materialCost_HDTV = unitCosts(1,1) * productionVolume(1,1);
```

Similarly, here is the material cost for Quarter 1 for the regular TV line:

```
>> q1_materialCost_RegTV = unitCosts(2,1) * productionVolume(2,1);
```

Do you see a regularity in how these scalar expressions are working out? Can you predict what the scalar expression would be to compute the Quarter 1 material cost for the M2 Computer line (the third row in the two data tables)?

The scalar expressions in the last paragraph are doing little more than you could have done to solve this problem with a calculator. But suppose we use MATLAB to compute all the material costs for Quarter 1. That is, we want to compute a vector that will tell us the material cost for each of the four product lines for Quarter 1. This computed vector will have four elements corresponding to the four product lines. At this point you should be thinking cell-by-cell multiplication. Why? Think this through.

Here is a MATLAB command to accomplish the desired computation:

```
>> q1_materialCost = unitCosts(:,1) .* productionVolume(:,1)
q1_materialCost =
     474544.00
      64219.00
     212100.00
      47526.00
```

We pull out the first column from **unitCost**, the first column in **productionVolume**, and apply the cell-by-cell multiplication operator to the two column vectors. The result is a column vector whose elements are the material costs for each product category for Quarter 1.

We could in a similar manner, compute the material costs for all product lines for Quarter 2, for Quarter 3, and for Quarter 4:

```
>> q2_materialCost = unitCosts(:,1) .* productionVolume(:,2);

>> q3_materialCost = unitCosts(:,1) .* productionVolume(:,3);

>> q4_materialCost = unitCosts(:,1) .* productionVolume(:,4);
```

The first column of **unitCosts** is used for each quarter because what we are computing for each quarter is the material costs, and the first column of **unitCosts** is the column for material costs. But the column of **productionVolume** varies: For the Quarter 1 result, we use the first column of **productionVolume**; for the Quarter 2 result, we use the second column; etc., because the **productionVolume** array is organized with quarter-specific data in its columns.

MATLAB allows us to be more efficient in carrying out the computations to tell us the material costs for all four quarters. What we would like to do is to perform one cell-by-cell multiplication that would produce an array that would be identical to the following:

```
[q1_materialCost,q2_materialCost,q3_materialCost, q4_materialCost]
```

We need to have the first column from **unitCosts** multiplied (cell-by-cell) with each column of **productionVolume**. The following MATLAB code accomplishes this task:

```
>> matCostCol = unitCosts(:,1);
>> matCostArray = [matCostCol, matCostCol, matCostCol, matCostCol];
>> allQuarters_materialCost = matCostArray .* productionVolume
allQuarters_materialCost =
```

474544.00	561068.00	606560.00	714492.00
64219.00	49021.00	57663.00	77778.00
212100.00	91175.00	146825.00	172200.00
47526.00	48505.00	44322.00	51353.00

Be sure you understand what this array **allQuarters_materialCost** represents, i.e., how to read the table. For example, you should be able to solve the material costs for the M2 Computer product in Quarter 3. Be sure you understand the logic that leads to a single MATLAB command being able to produce the array **allQuarters_materialCost**.

Suppose we examine the problem specification we are supposed to be working on. Your goal is to compute (a) the total costs for materials used on all four product lines for each quarter and (b) the total yearly cost for materials used in each of four product lines.

Each row in **allQuarters_materialCost** represents data for a given product line of the ABC factory, and each column represents data for one quarter of the last year. Part (a) requires a sum of each column of **allQuarters_materialCost**:

```
>> allQ_matCosts_totalByQuarter = sum(allQuarters_materialCost)
allQ_matCosts_totalByQuarter =
            798389.00        749769.00        855370.00      1015823.00
```

The result is a row vector. When working with arrays, you have to understand what the result of a cell-by-cell operation means. In this case, you can look back to the array for **allQuarters_materialCost**, and understand that the first element in the result **allQ_matCosts_totalByQuarter** ($798,389) is the total cost for materials for Quarter 1, the second element is for the total cost for materials for Quarter 2, the third element is for the total cost for materials for Quarter 3, and the last element is for the total cost for materials for Quarter 4. Be certain you understand this point.

Part (b) of the problem statement calls for a sum across rows:

```
>> allQ_matCosts_totalByProductLine = sum(allQuarters_materialCost,2)
allQ_matCosts_totalByProductLine =
    2356664.00
     248681.00
     622300.00
     191706.00
```

In this case, the result is expressed as a column vector, and you must understand what the elements of the result represent. The first element ($2,356,664) is the total yearly cost of materials for the HDTV line, and so on for corresponding values for the other three product lines produced by the ABC factory.

The mechanics of cell-by-cell operations are easy. The hard part is applying the capabilities of array cell-by-cell operations in a productive way and understanding the results. In the example above, you are asked to calculate two sets of sums: in Part (a), sums for total material costs for each quarter; and in Part (b) sums for the entire calendar year for each product line of ABC. The key step in the MATLAB solution is to realize that a single array (**allQuarters_materialCost**) can be used to represent material cost contributions for each product line for each quarter. Then you would see that summing down the columns yields the answer to Part (a) and summing across the rows yields the answer to Part (b).

As an exercise, develop MATLAB solutions to solve the following two variations on the example we just solved:

1. You must compute (a) the total costs for labor used on all four product lines for each quarter and (b) the total yearly cost for labor used in each of four product lines.

2. You must compute (a) the total costs for transportation used on all four product lines for each quarter and (b) the total yearly cost for transportation used in each of four product lines.

(You will find it easy to do the two variations above if you understood the solution to the example on materials costs. If you find it difficult to develop the MATLAB code to solve the two variations, go back and work through the solution for the entire example for materials costs again.)

MATLAB can represent and manipulate array data well. But for it to be useful, you must understand how to apply its capabilities to problems. We are not done with the example of the ABC factory yet. In the next section, you will learn about the other major class of arithmetic operations that can be applied to arrays: matrix operations. After you learn the basics of matrix operations, we will return to the ABC factory data and answer some more questions about it.

7-4.2.2   Matrix Operations on Arrays

In this section, you will learn to use and apply the matrix operations that are the last four entries in Table 7-1 and shown in isolation in Table 7-4. We will examine two of the four entries in Table 7-4: matrix multiplication of arrays (*) and matrix left division of arrays (\). The reason for leaving out matrix right division and matrix exponentiation is that full discussion of these operators would take us beyond the scope of this book and into the heart of matrix algebra.

Table 7-4: Arithmetic Operator Class for Arrays: Matrix Operators Only

Type	Operator name	Operator Symbol
matrix	multiplication	*
matrix	right division	/
matrix	left division	\
matrix	exponentiation	^

The matrix operations in Table 7-4 are rooted in an entire area of mathematics called matrix algebra. Many basic science students and engineering students will take a course in matrix algebra as a part of their major curricula, but often not until their junior year. In this book, we will go over the use of two of the matrix operators that MATLAB offers, but we will not assume you already have formal experience with matrix algebra. So, you will have to simply accept some of the following operations.

We will emphasize two types of applications of matrix operations in this section: (a) general word problems in which data can be represented in arrays and which can be solved economically by matrix multiply operations, and (b) problems represented by a set of simultaneous linear equations and solved using the matrix left division operation.

7-4.2.2.1   The Matrix Multiply Operation

Now you must learn what matrix multiplication is, i.e., how it is computed. Understanding the matrix multiplication operation at this level is a key to recognizing when matrix multiplication in MATLAB can be applied to a given problem.

The matrix multiplication operation C=A*B is defined by specifying a way to compute each element of C: for every element C(i,j) in C. The following relation defines how to calculate the value of the (*i,j*) element of C:

$$C(i, j) = \sum_{k} A(i, k) \times B(k, j) \qquad \text{(EQ 7-1)}$$

From this definition, three rules follow about the shape (number of rows and number of columns) of **A**, **B**, and **C**:

1. The number of columns in **A** must be equal to the number of rows in **B**. If this rule is not met, then the matrix multiplication operation between **A** and **B** is not a legal operation. This rule follows because a single *k* is used in the sum indicated in Equation 7-1 for the column index of **A** and for the row index of **B**.

2. Assuming Rule 1 is met (that is, assuming **A*B** is a legal operation) the number of rows in **C** will be equal to the number of rows in **A**.

3. Likewise, the number of columns in **C** will be equal to the number of columns in **B**.

Like most new subjects, the easiest way to understand the matrix multiplication operation is by understanding examples. As a first example, suppose we define a row vector **x** and a column vector **y** as follows:

```
>> x = [1, 2 3]
x =
            1.00              2.00              3.00
>> y = [10; 11; 12]
y =
          10.00
          11.00
          12.00
```

We compute **C=x*y**. First, is **x*y** legal? The shape of **x** is 1x3 (1 row, 3 columns), while **y** is a 3x1 array. Thus, Rule 1 above is satisfied. Second, what will be the shape of **x*y**? Since **x** is 1x3, and **y** is 3x1 **x*y** will have 1 row (by Rule 2 above) and 1 column (by Rule 3 above); therefore, **x*y** will be a 1x1 array.

Thus, only one application of Equation 7-1 needs to be made to compute all the elements of **C** since only one element of **C** exists.

$$C(1, 1) = \sum_k x(1, k) \times y(k, 1)$$

$$C(1, 1) = (x(1, 1) \times y(1, 1)) + (x(1, 2) \times y(2, 1)) + (x(1, 3) \times y(3, 1)) \qquad \textbf{(EQ 7-2)}$$

$$C(1, 1) = (1 \times 10) + (2 \times 11) + (3 \times 12)$$

$$C(1, 1) = 68$$

Figure 7-1 shows the computation of Equation 7-2 graphically. From a purely mechanical viewpoint, you are going across **A** to get successive terms to use in the running sum, while you are simultaneously going down **B** to get corresponding successive terms.

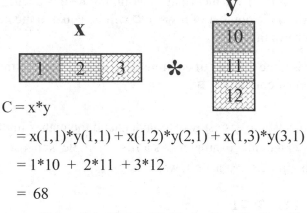

$$C = x*y$$

$$= x(1,1)*y(1,1) + x(1,2)*y(2,1) + x(1,3)*y(3,1)$$

$$= 1*10 + 2*11 + 3*12$$

$$= 68$$

**Figure 7-1: Matrix Multiplication of a Row Vector
Times a Column Vector**

The matrix multiplication operation is not the same as a scalar multiplication operation or as a cell-by-cell multiplication operation. One general way that matrix multiplication is different is that the order of the arrays to be multiplied makes a difference. For any two scalar numbers **s** and **t**, **st = ts**. In mathematics, this property that states order does not matter is called the commutative rule: You say that scalar multiplication is commutative. You should be able to see that cell-by-cell multiplication is also commutative.

Matrix multiplication, on the other hand, is not commutative: Order does make a difference. Look back at the specific **x** and **y** that we used in the example above. **x*y** in that example was the 1x1 array with the single element 68. But **y*x** is not a legal matrix multiplication operation. Hence, for this example, the commutative law does not hold. Since commutation does not hold in this one example, it cannot hold in general for the matrix multiplication operation.

As a second example of the matrix multiplication operation, consider two arrays **E** and **F**:

```
>> E = [1, 2, 3;  4, 5, 6]
E =

              1.00                 2.00                3.00
              4.00                 5.00                6.00
>> F = [10, 20;  30, 40;  50, 60]
F =
             10.00                20.00
             30.00                40.00
             50.00                60.00
```

E is a 2x3 array, and **F** is a 3x2 array. We want to compute **G=E*F**.

First, is **E*F** legal? Because the number of columns in **E** is equal to the number of rows in **F**, **E*F** is a legal matrix multiplication operation according to the first rule. Second, what shape will **G** have? **G** will have two rows (Rule 2) because **E** has two rows, and **G** will have 2 columns (Rule 3) because **F** has two columns. Therefore, **G** will be a 2x2 array. Third, how many times will we have to apply Equation 7-1 to compute all the element values in **G**? We have to apply Equation 7-1 to get a value for each element of **G**. Specifically, the number of elements in **G** is four. Determine for yourself that this is true, and be sure you know how to find the total number of elements in an NxM array.

The computation to get each one of the elements of **G=E*F** for two arrays **E** and **F** is analogous to the computation we did above in the case where there was only one element in **G**, i.e., in the case where **E** was a row vector and **F** was a column vector. This follows from applying the definition of matrix multiplication in Equation 7-1 and is shown graphically in Figure 7-2 for the (2,2) element of **G=E*F**.

Test your understanding of the matrix multiplication operation by solving the other three elements of **G=E*F** by hand. Compare your answers by performing the following from the MATLAB command window:

```
>> E   = [1, 2, 3;  4, 5, 6];
>> F = [10, 20;  30, 40;  50, 60];
>> G = E*F
G =
            220.00               280.00
            490.00               640.00
```

**Figure 7-2: Matrix Multiplication of Two Arrays for the
Row=2, Column=2 Element**

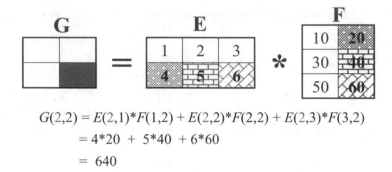

$$G(2,2) = E(2,1)*F(1,2) + E(2,2)*F(2,2) + E(2,3)*F(3,2)$$
$$= 4*20 + 5*40 + 6*60$$
$$= 640$$

You are probably asking yourself why we have gone over the matrix multiplication operation in such detail. After all, matrix multiplication is available as an operation in MATLAB, so why do you need to know how to perform matrix multiplication by hand? The answer is that if you don't know how matrix multiplication works, then you will have a hard time knowing when to apply the appropriate tool.

Compare what happens when you perform a cell-by-cell multiplication to two arrays versus what happens when you apply a matrix multiplication to two arrays. As you know by now, cell-by-cell multiplication performs a scalar multiplication between corresponding elements of two arrays. But as you have learned, matrix multiplication performs a series of scalar multiplications and keeps a running sum of the products. Many times a problem specification will ask you for a result that is a sum of a number of terms. When using MATLAB, whenever you face such a problem, you should consider representing the problem using MATLAB arrays and explore the possibility that matrix multiplication may be the tool of choice.

Suppose we return to the problem of costs of materials, labor, and transportation at the ABC electronics firm. Initially, we were given two data tables for the ABC company, as reproduced in Table 7-5 and Table 7-6. Here is the problem we worked with earlier:

> *Your goal is to compute (a) the total cost for materials used on all four product lines for each quarter and (b) the total yearly cost for materials used in each of four product lines.*

We solved this example by applying cell-by-cell operations, in particular by using cell-by-cell multiplication. Suppose we consider broadening the problem:

**Table 7-5: ABC Production Costs for
Each Unit Produced**

Product	Material Cost ($)	Labor Cost ($)	Transportation Cost ($)
48-in HDTV	892	531	89
32-inch TV	149	128	15
M2 computer	175	75	15
R2 DVD	89	22	5

Table 7-6: ABC Production Volume Last Year

Product	Quarter 1	Quarter 2	Quarter 3	Quarter 4
48-inch HDTV	532	629	680	801
32-inch TV	431	329	387	522
M2 computer	1212	521	839	984
R2 DVD	534	545	498	577

*Your goal is to compute (a) the total
costs for materials, labor, and transportation for each quarter
(sum of all product lines) and (b) the total yearly cost for materials, labor, and transportation.*

Start out by representing the data tables in Table 7-5 and Table 7-6 as MATLAB arrays:

```
>> unitCosts = [892, 531, 89; ...
                149, 128, 15; ...
                175,  75, 15; ...
                 89,  22,  5];
>> productionVolume = [532, 629, 680, 801; ...
                       431, 329, 387, 522; ...
                       1212, 521, 839, 984; ...
                       534, 545, 498, 577];
```

Now, determine what you need to solve the problem. You are asked to figure out the costs for each quarter for ABC from all its product lines for each of the three cost categories. This would answer Part (a) in the

Table 7-7: Form of the Target for Revised Problem Spec

	Q1	Q2	Q3	Q4
Total Materials	??	??	??	??
Total Labor	??	??	??	??
Total Transportation	??	??	??	??

revised problem. When you see a problem like this, a good way to begin is to sketch what the target (the result) of the computation will look like. In our example, one form the result could take is depicted in Table 7-7. (Another would be the transpose of the data table in Table 7-7.)

Suppose we focus on the (1,1) element of the data table in Table 7-7: the total materials cost for Quarter 1. With paper and pencil, you could calculate this value as:

$$totalMaterialCostsQ1 = materialCostHDTVQ1 + materialCostRegTVQ + \ldots$$
$$materialCostComputerQ1 + materialCostDVDQ1$$

Each of these terms is the unit cost for the item times the number of items manufactured, e.g.,

$$materialCostHDTVQ1 = unitCostHDTV \times numHDTVinQ1$$

Four terms are to be summed in **totalMaterialCostsQ1**, and each term is a product. This form is what the matrix multiplication operation does. So, the solution to Part (a) may be a simple application of matrix multiplication.

We could compute the value of the (1,1) element of data table Table 7-7:

```
>> totalMaterialCostsQ1 = sum(unitCosts(:,1) .* productionVolume(:,1))
totalMaterialCostsQ1 =
      798389.00
```

Then we would be left to perform a similar computation for each of the other eleven elements of Table 7-7. On the other hand, we could compute the value of the (1,1) element of data table Table 7-7 as follows:

```
>> totalMaterialCostsQ1 = unitCosts(:,1)' * productionVolume(:,1)
totalMaterialCostsQ1 =
      798389.00
```

To arrange the computation to perform the sum of products as desired using matrix multiplication, we needed to take the transpose of the vector coming from **unitCosts**.

The two ways to compute the value of the (1,1) element of data table Table 7-7 are equivalent (they give the same result), but the second variation using matrix multiplication can be extended to give us all the other eleven elements in Table 7-7 with no additional work:

```
>> allCosts_AllQuarters = unitCosts' * productionVolume
allCosts AllQuarters =
       798389.00    749769.00    855370.00   1015823.00
       440308.00    427176.00    484497.00    578641.00
        71663.00     71456.00     81400.00     96764.00
```

Confirm that, for example, the (3,2) element in **allCosts_AllQuarters** is correct. This (3,2) element should contain a value that is the total transportation costs for Quarter 2.

To answer Part (b) of the revised problem (total yearly costs), we can apply the built-in **sum** function, directing it to sum across rows:

```
>> allCosts_YearlyTotals = sum(allCosts_AllQuarters,2)
allCosts_YearlyTotals =
      3419351.00
      1930622.00
       324283.00
```

We have talked at length about this problem of ABC Electronics costs and production. But if you go back and distill the previous MATLAB operations in the sections on cell-by-cell multiplication and matrix multiplication, you will see that once you represent data tables like the costs and production volumes of ABC in MATLAB, you can perform a tremendous variety of computations with the data tables with little MATLAB code construction.

The key to effective use of MATLAB does not lie in code construction. Rather, your first and most challenging goal is to represent data in MATLAB effectively and to think through which operations of MATLAB are the best to apply to any given problem. Once those harder steps are accomplished, the appropriate MATLAB code is relatively easy to build.

7-4.2.2.2   Matrix Left Division

In the discussion of the matrix multiplication operation, you learned that matrix multiplication is a very useful MATLAB operation, provided you can effectively map a problem into a solution using matrix multiplication. In this section, application of the matrix left division will be specifically described for one operation. Care must still be exercised to apply matrix left division properly, but the type of problem we will consider is less broad. In this section, we will show the use of matrix left division as a tool to solve mathematical systems of linear simultaneous equations.

Suppose you want to solve the system of linear simultaneous equations shown in Equation 7-3:

$$
\begin{aligned}
3x + 4y + 5z &= 32 \\
21x + 5y + 2z &= 20 \\
x - 2y + 10z &= 120
\end{aligned}
\qquad \text{(EQ 7-3)}
$$

These equations have three unknowns: x, y, and z. The goal is to find values of x, y, and z that make the three equations yield the values given on the right side of each equation.

Remembering your knowledge of the matrix multiplication operation, the three equations above may be written as a matrix product:

$$
\begin{bmatrix} 3 & 4 & 5 \\ 21 & 5 & 2 \\ 1 & -2 & 10 \end{bmatrix} \times \begin{bmatrix} x \\ y \\ z \end{bmatrix} = \begin{bmatrix} 32 \\ 20 \\ 120 \end{bmatrix}
\qquad \text{(EQ 7-4)}
$$

Verify for yourself by hand that Equation 7-4 follows from Equation 7-3.

Make the following variable assignments:

$$coefficientsArray = \begin{bmatrix} 3 & 4 & 5 \\ 21 & 5 & 2 \\ 1 & -2 & 10 \end{bmatrix}$$

$$unknownsVector = \begin{bmatrix} x \\ y \\ z \end{bmatrix} \qquad \text{(EQ 7-5)}$$

$$rightSideVector = \begin{bmatrix} 32 \\ 20 \\ 120 \end{bmatrix}$$

Equation 7-4 can be written in concise form:

$$\text{oefficientsArray} \times \text{unknownsVector} = \text{rightSideVecto} \qquad \text{(EQ 7-6)}$$

Equation 7-6 is an example of a matrix algebra equation. Using matrix left division, Equation 7-6 is solved for the **unknownsVector** in one operation. Before we turn to that one-step solution though, there is an important intuition that you need to build

If you had the following algebra equation $2x = 10$, you would find x by dividing both sides of the algebra equation by the scalar two. That would yield $x = 10/2 = 5$. This amounts to multiplying $2x = 10$ by the inverse of two: `2x*(1/2) = 10*(1/2)`, which results in $x = 5$.

But when you have a matrix algebra equation like **Ax = y**, where **A** is an array and **x** and **y** are vectors, more care has to be used to solve for the **x** vector. Although a full discussion of solving **Ax = y** is well beyond the scope of this book, there is one point you know that will help you understand the need for extra care. Recall that for two arrays **A1** and **A2**, matrix multiplication is not commutative: **A1*A2** is not necessarily equal to **A2*A1**. To solve **Ax = y**, you must multiply both sides of the matrix equation by the inverse of **A**, and you must apply the operation as `(1/A)*A*x = (1/A)*y`. The order on both sides of the equation is important because matrix multiplication is not commutative. This is the reason that MATLAB provides a left and a right matrix multiplication operation.

To solve the matrix algebra equation of Equation 7-6 for the **unknownsVector**, the MATLAB left matrix division will do the trick[4]:

$$\text{nknownsVector} = \text{coefficientsArray} \setminus \text{rightSideVecto} \qquad \text{(EQ 7-7)}$$

To solve Equation 7-3 for **x**, **y**, and **z**, the following MATLAB code can be used:

```
>> coefficientsArray = [ 3,   4,   5; ...
                        21,   5,   2;...
                         1,  -2,  10];
>> rightSideVector = [ 32;...
                       20;...
                      120];
>> unknownsVector = coefficientsArray \ rightSideVector
unknownsVector =
          1.45
         -6.32
         10.59
```

That is, **x = 1.45**, **y = -6.32**, and **z = 10.59**.

There are two important cautions on the use of matrix operations to solve systems of linear simultaneous equations. The first is that some systems of linear equations do not have analytical solutions. There is good news and bad news here. The bad news is you cannot solve a system of linear simultaneous equations for some situations, e.g., matrix left division will not work in some cases. Some situations do not have a solution, no matter what method you apply.

Consider the following set of simultaneous linear equations:

$$x - y = 32$$
$$x - z = 10 \qquad \text{(EQ 7-8)}$$
$$y - z = 22$$

Formulate these three equation in matrix algebra form:

$$\begin{bmatrix} 1 & -1 & 0 \\ 1 & 0 & -1 \\ 0 & 1 & -1 \end{bmatrix} \times \begin{bmatrix} x \\ y \\ z \end{bmatrix} = \begin{bmatrix} 32 \\ 10 \\ 22 \end{bmatrix} \qquad \text{(EQ 7-9)}$$

Set up these equations for solution using linear algebra in MATLAB:

```
>> coefficientsArray = [ 1,  -1,   0; ...
                         1,   0,  -1;...
                         0,   1,  -1];
```

4. MATLAB provides an inverse matrix operation: **inv**. To solve **Ax = y**, we could use **x = inv(A)*b**. But in MATLAB, use the matrix left division as shown in this section. The direct reasons are that using matrix left division for solving systems of linear simultaneous equations will be faster and provide more accurate results, particularly for large systems.

```
>> rightSideVector = [ 32;...
                      10;...
                      22];
```

Solve for the unknowns using matrix left division. MATLAB will hand you a warning as follows:

```
>> unknownsVector = coefficientsArray \ rightSideVector
Warning: Matrix is singular to working precision.
(Type "warning off MATLAB:singularMatrix" to suppress this warning.)
unknownsVector =
   Inf
   Inf
   Inf
```

Though it appears you have three equations in three unknowns in the system, that is not really the case. The situation is analogous to the simple algebra equation:

$$3x + 2 = (5x + 39) - (2x + 10)$$

If you try to solve this simple relationship for x, you will discover that you cannot find a unique x. The warning MATLAB gave indicating *Matrix is singular* is a tip that the system of linear simultaneous equations you want to solve has no unique solution. Either the physical system you are modeling must be dealt with in a numerical modeling method, i.e., not an analytic method, or you have made a mistake in representing your system of equations.

The good news is that linear algebra gives you a way to check if there is a solution by providing a way to check if a matrix is singular. And MATLAB implements what linear algebra gives. The check is to compute the *determinant* of the matrix. A full discussion of what the determinant operation really does is beyond our discussion here, but it is important for you to know that to compute the determinant of an array using MATLAB, you can use the **det** built-in function. If the determinant of an array is equal to zero, then the array is singular. For example, when MATLAB issued the warning above, you can check the array using **det**:

```
>> coefficientsArray = [ 1, -1,  0; ...
                         1,  0, -1;...
                         0,  1, -1];
>> det(coefficientsArray)
ans =
      0
```

The first caution about the use of matrix algebra is embedded in the mathematics itself. As long as you know the possibility that a system of linear simultaneous equation may not have a direct mathematical solution, you can at least detect the difficulty in a specific problem by using the **det** function.

The second caution is more subtle, and like many situations in which you apply mathematics to solve a science or engineering problem, the red flags start to come out only when you understand the physical problem you want to solve.

Suppose you are given the following problem:

> Jeanie, Juan, and Alexander each have some fruit. Each has a number of apples, oranges, and pears. All apples have the same weight, all oranges have the same weight, and all pears have the same weight. Jeanie has 3 apples, 2 oranges, and 1 pear. The total weight of fruit that Jeanie has is 52 ounces. Juan has 2 apples, 3 oranges, and 1 pear. The total weight of fruit that Juan has is 50 ounces. Alexander has 1 apple, 2 oranges, and 3 pears. The total weight of fruit that Alexander has is 56 ounces.

> What is the weight of each apple, orange, and pear?

Here is a matrix algebra solution using MATLAB for this problem:

```
>> fruitNumbersCoef = [3 2 1;
                       2 3 1;
                       1 2 3];
>> totalPoundsEach = [52;
                      50;
                      56];
>> weightEachFruit = fruitNumbersCoef \ totalPoundsEach
weightEachFruit =
    9.00
    7.00
   11.00
```

So far so good. The solution is that apples weigh 9 ounces, oranges weigh 7 ounces, and pears weigh 11 ounces.

But suppose we changed the problem statement so that Juan has ten pears instead of one as in the first problem statement:

> Now, what is the weight of each apple, orange, and pear?

A matrix algebra solution using MATLAB for the altered problem might appear to be:

```
>> fruitNumbersCoef = [3 2 1;
                       2 3 10;   % note change in (2,3) element
                       1 2 3];
>> totalPoundsEach = [52;
                      50;
                      56];
>> weightEachFruit = fruitNumbersCoef \ totalPoundsEach
weightEachFruit =
   -7.50
   40.00
   -5.50
```

Applying the mathematics of linear algebra here is easy; **fruitNumbersCoef** is not singular, and the matrix left division operation produces a perfectly valid result mathematically. So what is wrong? Look at the results again. The results state that apples weigh –7.5 ounces, oranges weigh 40 ounces, and pears weigh –5.5 ounces. But in the physical world, fruit can not have a negative weight. So, the problem as formulated has *no solution* in the real physical world.

To summarize, the two cautions about the use of matrix algebra to solve systems of linear simultaneous equations are that if the involved coefficients array is singular, no analytical unique solution is possible and you must pay attention to the results of the computation to see if the results are physically realistic.

The left division matrix operator is important to master because of its utility in solving systems of simultaneous linear equations. Many times, in real-world engineering situations, complex physical systems can be modeled (approximated) by systems of linear equations. This is particularly true where modeling complex mechanical systems is often accomplished by developing linear models that can be solved using the MATLAB left division matrix operator.

7-4.2.3   Array Relational Operators and Array Logic Operators, and Array Logical Built-in Functions

Array relational and logic operators act as cell-by-cell operators, so everything you learned earlier about vector relational and logic operators holds true for array relational and logic operators. To apply array relational and logic operators to two arrays **A** and **B**, **A** and **B** must have the same shape: same number of rows and same number of columns.

Suppose we define two arrays A and B:

```
>> A = [1, 2; ...
        3, 4; ...
        5, 6];
>> B = [6, 5; ...
        4, 3; ...
        2, 1];
```

We want to test for elements in **A** greater than the corresponding elements in **B**:

```
>> test_aGREATERb = A > B
test_aGREATERb =
              0                    0
              0                 1.00
           1.00                 1.00
```

The zeroes in the corresponding result array **test_aGREATERb** mean that the test returned **FALSE** for those array elements in **A** and **B**, and the ones indicate the test returned **TRUE**. In the above example, the relational operator **>** acts as a cell-by-cell operator. This point generalizes to all the relational operators as well as the logic operators when applied to arrays. Put another way, there is nothing new to learn about relational and logic operators applied to vectors beyond what you know about application to vectors.

In our earlier discussion on using the relational and logic operators with vectors, we described the use of the **find** built-in function to yield the locations of vector elements at which some test is **TRUE**. Extending the use of **find** to arrays requires you to stretch a bit beyond what you already know. For vectors, we need one index in a given vector to locate where that element is in a vector. Hence, **find** returned one vector of index values where the given test is **TRUE**:

```
>> aVector = [1, 2, 3, 4, 5]
aVector =
          1.00      2.00      3.00      4.00      5.00
>> locations_aVectorGreater3 = find(aVector > 3)
locations_aVectorGreater3 =
          4.00      5.00
```

If you are dealing with an array test and want to use **find** to locate positions in the array where a test is **TRUE**, then you will need two index values to locate each position in the array where the test is **TRUE**. For example, the same **A** and **B** arrays:

```
>> A = [1, 2; ...
        3, 4; ...
        5, 6];
```

```
>> B = [6, 5; ...
        4, 3; ...
        2, 1];
>> [rowTestTrue, columnTestTrue] = find(A > B)
rowTestTrue =
            3.00
            2.00
            3.00
columnTestTrue =
            1.00
            2.00
            2.00
```

find applied to arrays returns two output arguments: a vector of row locations and a vector of column locations. The two returned outputs correspond to row and column locations in the test arrays. So, for example, one location in **A** and **B** where **A>B** is the location **(rowTestTrue(1),columnTestTrue(1))**. The complete set of **(row, column)** locations where the test is **TRUE** can be found:

```
>> locationsTestTrue = [rowTestTrue, columnTestTrue]
locationsTestTrue =
            3.00            1.00
            2.00            2.00
            3.00            2.00
```

The values in the first column of **locationsTestTrue** are row index values, and the values in the second column of **locationsTestTrue** are the corresponding column index values. As long as you understand that to locate an element in an array you need two index values, one for row location and one for column location, the use of **find** applied to arrays will be straightforward.

Finally, there is one more thing to learn about using **find**: how to extract *values* in an array that meet some logical test. Suppose we have an array **D**:

```
>> D = [2, 4; 5, 3];
```

We want to extract all the values in **D** that are, for example, greater than three. Above you saw how to extract the *locations* of cells in **D** that meet some test; now you want to get the *values* out. To perform the operation, you use **find** as an index into **D** as follows:

```
>> valuesInD_greater3 = D(find(D>3))
valuesInD_greater3 =
        5
        4
```

The results are returned as a column vector. To find locations the above method will return a two-column array, with its number of rows corresponding to the rows in the column vector of values you learned how to extract.

7-4.3   Array Expressions and Rules for Forming Array Expressions

In chapter five, in the section on vector expressions and rules for forming vector expressions (Section 5-4.3), we showed a table with the operator precedence for all cell-by-cell scalar and vector operations (Table 5-7). In this chapter you have extended your understanding of operators to include array operators, and you have seen that the application to all cell-by-cell operators is straightforward. In addition, you have also learned about matrix operations, matrix multiplication and matrix left division in particular. To be complete, Table 7-8 folds these new operators (and matrix exponentiation and matrix right division) into a complete listing for the precedence of MATLAB operators.[5] Remember, the purpose of the precedence table of Table 5-7 is to tell you the order that MATLAB will use in processing an expression: Higher precedence level values indicate priority in the order that MATLAB uses to execute an expression. The parentheses operator trumps the precedence level of all other operators. So, you can always make it clear what your intent is when you write an array expression by using parentheses.

7-4.4   Section Synopsis

This section, and this whole chapter, has been aimed at teaching you to represent problems in MATLAB arrays and how to operate on MATLAB arrays.

Although the material in this chapter is among the most important in this entire book because it is a culmination of earlier chapters, there is a really short summary of material in this chapter. Cell-by-cell operations on MATLAB arrays is an easy extension from cell-by-cell operations on vectors; the key matrix operations that are new are matrix multiplication and matrix left division.

---

5. We have omitted two MATLAB operators: short circuit AND (**&&**) and short circuit OR (**||**). These two operators behave conceptually like AND and OR and are in the complete MATLAB set of operators for increased program speed. See MATLAB Help for more details on **&&** and **||** .

Table 7-8: Precedence Rules for MATLAB Operators

Precedence Level	Operator
1	parentheses
2	transpose (**'**) cell-by-cell exponentiation (**.** $^\wedge$) matrix exponentiation ($^\wedge$)
3	cell-by-cell unary plus (**+**) cell-by-cell unary minus (**−**) logical negation (**~**)
4	cell-by-cell multiplication (**.** $*$) matrix multiplication ($*$) cell-by-cell right division (**.** **/**) matrix right division (**/**) cell-by-cell left division (**.** ****) matrix left division (****)
5	cell-by-cell addition (**+**) cell-by-cell subtraction (**−**)
6	the colon operator (**:**)
7	cell-by-cell relational *all the relational operators* (**<, <=, >, >=, ==, ~=**)
8	cell-by-cell logical AND (**&**)
9	cell-by-cell logical OR (**|**)

1. Arrays are indexed by giving two index values: one for row location and one for column location.

2. All array accesses are a generalization of what you know for vector access with the only specific difference being that you need to use two index values for any array access.

3. All cell-by-cell operations are generalizations of the corresponding vector operation.

4. Matrix multiplication can be very economical (in the amount of code you have to write) when the problem you are solving at its root involves a sum of a number of scalar multiplication operations. The whole key to applying vector multiplication appropriately lies in understanding the problem you are solving and looking for opportunities where matrix multiplication might be applied.

5. Matrix left division is often used to solve systems of linear simultaneous equations. Use the `det` function to determine if a matrix left division is mathematically possible. As always, be careful to interpret the results you obtain with MATLAB.

6. Care must be exercised when the `find` built-in function is applied to an array. Two output values are returned: a vector of row index values where an indicated relational/logical test is True and a corresponding vector of column index values where the test is true.

7. Values in an array that meet some relational test may be extracted from the array using `find` as an indexing term.

Synopsis for Section 7-4

7-5   Problem Sets for Arrays

7-5.1   Set A: Nuts and Bolts Problems for Arrays

Problem 7-A.1 (Section 7-1)

Write a function, **Prob7_A_1**, of no inputs and one array output. Internally, your function should construct the following array using the square bracket operator and should output the array.

$$\begin{bmatrix} 1 & 4 & 3 & 7 & 2 \\ 6 & 8 & 9 & 5 & 1 \end{bmatrix}$$

Problem 7-A.2 (Section 7-1)

Write a function, **Prob7_A_2**, of no inputs and one array output. Internally, your function should construct the following array using the square bracket operator and should output the array.

$$\begin{bmatrix} 8 & 6 \\ 7 & 8 \\ 9 & 1 \\ 6 & 3 \\ 3 & 1 \end{bmatrix}$$

Problem 7-A.3 (Section 7-1 and Section 7-3)

Write a function, **Prob7_A_3**, of no inputs and two array outputs. Internally, your function should construct the following two arrays using the square bracket operator, the colon operator (or alternative to the colon operator, the **linspace** function), and the transpose operator. The two arrays constructed should be your functions outputs.

$$\text{arrayA} = \begin{bmatrix} 1 & 11 & 21 & 31 & 41 & 51 & 61 & 71 \\ 80 & 75 & 70 & 65 & 60 & 55 & 50 & 45 \\ 3 & 6 & 9 & 12 & 15 & 18 & 21 & 24 \\ 99 & 89 & 79 & 69 & 59 & 49 & 39 & 29 \end{bmatrix}, \quad \text{arrayB} = \begin{bmatrix} 1 & 80 & 3 & 99 \\ 11 & 75 & 6 & 89 \\ 21 & 70 & 9 & 79 \\ 31 & 65 & 12 & 69 \\ 41 & 60 & 15 & 59 \\ 51 & 55 & 18 & 49 \\ 61 & 50 & 21 & 39 \\ 71 & 45 & 24 & 29 \end{bmatrix}$$

Hint: Do you notice any patterns in the rows of **arrayA**? For example, the first row of **arrayA** could be created by **1:10:71** or **linspace(1,71,8)**. Once you figure out how to create each row, you can concatenate them.

Problem 7-A.4 (Section 7-1)

Write a function, **Prob7_A_4**, of no inputs and one array output. Internally, your function should construct the following array and should output it.

$$\begin{bmatrix} 7 & 6 & 44 & -9 & 61 \\ 14 & 4 & 55 & -5 & 36 \\ 21 & 2 & 66 & -1 & 11 \\ 28 & 0 & 77 & 3 & -14 \\ 35 & -2 & 88 & 7 & -39 \\ 42 & -4 & 99 & 11 & -64 \end{bmatrix}$$

Problem 7-A.5 (Section 7-1)

Write a function, **Prob7_A_5**, of no inputs and one array output. Internally, your function should construct the following array and should output it.

$$\begin{bmatrix} 0 & 0 & 0 \\ 8 & 8 & 8 \\ -32 & -32 & -32 \end{bmatrix}$$

Hint: To create the second row, we could use **8*ones(1,3)**.

Problem 7-A.6 (Section 7-1)

Write a function, **Prob7_A_6**, of no inputs and one array output. Internally, your function should construct the following array and should output it.

$$\begin{bmatrix} 4 & 4 & 4 & 4 & 0 & 0 & 0 & 0 \\ 6 & 4 & 2 & 0 & 41 & 51 & 61 & 71 \\ -5 & -10 & -15 & 16 & 18 & 20 & 22 & 24 \\ 7 & 14 & 21 & 28 & 21 & 14 & 7 & 0 \end{bmatrix}$$

Problem 7-A.7 (Section 7-2)

Write a function, **Prob7_A_7**, that takes one array input and produces one array output. Internally, your function should construct the single output array with elements taken from the input array, specifically from the second and third rows and from the first three columns of the input array. Assume that the input array has dimensions n x m such that $n > 3$, and $m > 3$.

What you are doing is pulling out a number of specified elements for construction of the output from the input.

Problem 7-A.8 (Section 7-2)

Write a function, **Prob7_A_8**, that takes one array input and produces one array output. Internally, your function should construct the single output array with elements taken from the input array, specifically from the third and fourth rows and from the last five columns of the input array. Assume that the input array has dimensions n x m such that $n > 4$, and $m > 5$. In your indexing solution, use **end**.

Problem 7-A.9 (Section 7-2)

Write a function, **Prob7_A_9**, that takes one array input and produces one array output. Internally, your function should set the last four elements in the first row of the input to 2 and return the modified array. Assume that the input array has dimensions n x m such that $m > 3$.

Problem 7-A.10 (Section 7-2)

Write a function, **Prob7_A_10**, that takes one array input and produces one array output. Internally, your function should set the four corners of the input array to 99 and return the modified array.

For example, if the input array has dimensions 2 x 3, then the modified array would be such that elements (1,1), (2,1), (1,3), and (2,3) would all have the value 99.

Problem 7-A.11 (Section 7-2)

Write a function, **Prob7_A_11**, that takes one array input and produces one vector output. Internally, your function should pull out the last row of the input array and return it as the vector output.

Test your function with the following inputs:

    [5] (note that the last row of scalar 5 is the scalar 5)

    [1 2 3 4 5]

    [1 2 3 4 5]'

    [1 2 3 4 5; 5 4 3 2 1]

Problem 7-A.12 (Section 7-2)

Write a function, **Prob7_A_12**, that takes one array input and produces one array output. Internally, your function should reverse the order of elements in the second column of the input and output the resultant modified input array. Assume that the input array has dimensions n x m such that $m > 1$.

Hint: A vector **V** can be reversed by **V = V(end:-1:1)**.

Problem 7-A.13 (Section 7-4)

For the two matrices below, find **A*B** by hand. Verify your answer using MATLAB.

$$A = \begin{bmatrix} 1 & 2 & 0 \\ 2 & 1 & 3 \end{bmatrix}, \quad B = \begin{bmatrix} 2 & 1 \\ 0 & 1 \\ 1 & 4 \end{bmatrix}$$

(Do the operation by hand. If you do not understand the concept of matrix multiplication, you are not going to understand when to apply it later in complex word problems.)

Problem 7-A.14 (Section 7-4)

Fill out the table to the right with a Yes or a No in the column marked Legal? indicating whether a matrix multiplication is legal for arrays of the stated dimensions.

Is **C=A*B** legal? If the operation is legal, then what are the dimesions of **C**?

# Rows in **A**	# Cols in **A**	# Rows in **B**	# Cols in **B**	Legal?	C Dimensions
2	3	3	3		
3	3	2	2		
2	3	2	3		
2	3	3	1		
10	12	12	2		

In the column marked **C Dimensions**, fill in the dimensions of the array resulting from the matrix multiplication if the operation is legal.

Problem 7-A.15 (Section 7-4)

For the matrices **A** and **B** below, does **A*B** give the same result as **B*A**?

$$A = \begin{bmatrix} 1 & 3 \\ 2 & 4 \end{bmatrix}, \quad B = \begin{bmatrix} 2 & 0 \\ 1 & 3 \end{bmatrix}$$

Work out your answer by hand first and then verify your result using MATLAB.

Problem 7-A.16 (Section 7-4)

Write a function, **Prob7_A_16**, that takes no input and returns a vector solving the following system of equations using matrix operations.

$$2x + 5y + z = 12$$
$$x + 2y - 3z = 5$$
$$x - y + z = -1$$

Problem 7-A.17 (Section 7-4)

Write a function, **Prob7_A_17**, that takes no input and returns a vector solving the following system of equations using matrix operations.

$$x + y - z = 6$$
$$2x - y + 2z = 3$$
$$3x + 2y + 5z = 35$$

Problem 7-A.18 (Section 7-4)

Write a function, **Prob7_A_18**, that solves a set of linear equations. Inputs to your function should be (a) a vector of the right-hand side values for system of linear equations and (b) an array of coefficients for the set of equations. The output should be a column vector with the solution for the unknowns in the system.

Problem 7-A.19 (Section 7-4)

Write a function, **Prob7_A_19**, that inputs one array, and produces the following output:

(a) An array containing the *locations* of elements in the input that are less than zero

(b) A corresponding column vector containing the *values* of elements in the input that are less than zero

Use **find** in your solution.

For example, if the input array is:

```
>> A = [3 -5 9; -22 8 7; -4 10 -11]
A =
      3    -5     9
    -22     8     7
     -4    10   -11
```

A call to **Prob7_A_19(A)** should produce the following:

```
>> [locations, values] = Prob7_A_19(A)
locations =
     2     1
     3     1
     1     2
     3     3
values =
   -22
    -4
    -5
   -11
```

So, each row of **locations** is an identification of a cell in the input that is less than zero, and the corresponding row in **values** contains the value in the input at that location.

Problem 7-A.20 (Section 7-4)

Write a function, **Prob7_A_20**, that takes two inputs: an array and a scalar. Your function should return two logic values: (a) a logic value holding **TRUE** if all elements of the input array are greater than the scalar input and holding **FALSE** if otherwise, and (b) a similar logic value but testing if any of the elements of the input array is greater than the scalar input.

For example, using the same input array as in the last problem:

```
>> A = [3 -5 9; -22 8 7; -4 10 -11];
```

A call to **Prob7_A_20(A,2)** should produce the following:

```
>> [allG, anyG] = Prob7_A_19(A, 2)
allG =
     0
anyG =
     1
```

Problem 7-A.21 (Section 7-4)

Write a function, **Prob7_A_21**, that inputs two arrays of the same dimensionality and outputs **TRUE** if every element of the first input array is greater than the corresponding element in the second input array.

Problem 7-A.22 (Section 7-4)

Write a function, **Prob7_A_22**, that inputs two arrays of the same dimensionality and outputs **TRUE** if every element of the first input array is equal to the corresponding element in the second input array.

7-5.2   Set B: Problem Solving with MATLAB Arrays

Problem 7-B.1

Write a function, **Prob7_B_1**, that takes a row vector of temperatures in degrees Fahrenheit and outputs a two-column array. The first column should consist of the input values in degrees Fahrenheit. The second column should contain the corresponding temperatures in degrees Celsius. Use Google if you do not remember the conversion relationship.

Problem 7-B.2

Your professor requests help with calculation of final grades and grade analysis. *(Dream on!)* He would like to have a function, **Prob7_B_2**, that will input a table of student numbers and student grades and output the final grade for each student, class grade average, and the student number of the top scoring student.

Student #	MT	HW	P	F
273	90	75	91	100
677	65	72	85	87
496	80	82	100	74
641	100	98	95	65
911	92	77	90	81
216	95	83	90	80

A sample input to your function is provided to the right. In this table, the columns contain the student numbers, grades for the midterm, homework, term project and the final exam. Your professor indicates that each student's weighted final grade should be calculated using the following equation:

$$\text{finalGrade} = 0.3MT + 0.2HW + 0.2P + 0.3F$$

Problem 7-B.3

Write a function, **Prob7_B_3**, that inputs an $N \times M$ array, finds row and column sums, and concatenates the row and column sums as the last column and row, respectively. The $(N+1, M+1)$ location should contain the sum of all data in the input array.

Your function should output the updated array. Here is an example input:

$$\begin{bmatrix} 2 & 1 & 5 \\ 4 & 6 & 2 \end{bmatrix}$$

Your output array should be as follows:

$$\begin{bmatrix} 2 & 1 & 5 & 8 \\ 4 & 6 & 2 & 12 \\ 6 & 7 & 7 & 20 \end{bmatrix}$$

Problem 7-B.4

The National Widget Company sells four types of widgets. Data in the following sample table shows the selling price and number of items sold for each widget in 2002, 2003, and 2004.

Product ID	Selling Price ($)	# Items Sold in 2002	# Items Sold in 2003	# Items Sold in 2004
Widget 1	75	162	157	141
Widget 2	36	421	456	504
Widget 3	78	64	66	75
Widget 4	94	49	56	60

Write a function, **Prob7_B_4**, that inputs (a) a column vector containing the selling price for each type of widget and (b) an array containing the yearly sales data for each type of widget. For example, for the data in the above table, here are the two inputs:

$$\mathbf{sellingPrice} = \begin{bmatrix} 75 \\ 36 \\ 78 \\ 94 \end{bmatrix}, \quad \mathbf{salesData} = \begin{bmatrix} 162 & 157 & 141 \\ 421 & 456 & 504 \\ 64 & 66 & 75 \\ 49 & 56 & 60 \end{bmatrix}$$

Your function should output (a) a row vector containing the total revenue for each year of data in **salesData**, and (b) a column vector containing the total revenue over all years for each widget. For example, for the data in the above table, the two outputs from **Prob7_B_4** should be as follows:

$$\mathbf{totalYearlyRevenue} = \begin{bmatrix} 36904 & 38603 & 40209 \end{bmatrix}$$

$$\mathbf{totalAllYearsEachWidgetType} = \begin{bmatrix} 34500 \\ 49716 \\ 15990 \\ 15510 \end{bmatrix}$$

Hint: You can compute the **totalYearlyRevenue** output using a matrix operation. But the most straightforward way to compute the **totalAllYearsEachWidgetType** output is not by using a matrix multiplication.

Problem 7-B.5

Suppose you have a catalogue of automobiles following the form of the table below.

ID #	Color (0=Black, 1=Blue, 2=Red)	Type (0=Sedan, 1=Coupe, 2=Convertible)	Price ($ x 10^3)
1002	0	1	24
2176	1	2	18
3201	2	1	36
4204	1	1	26
4110	2	2	35
5611	1	2	23
7556	0	0	29
8732	1	0	31
9666	0	1	28

Your goal is to find cars (IDs of cars) that meet the criteria of having a stated color, being of a specified type, and costing less than a specified amount.

Write a MATLAB function, **Prob7_B_5**, that will input the following:

(a) The numerical data in the format of table shown above

(b) Desired color of car (as a number: see the table above)

(c) Desired type of a car (as a number: see the table above)

(d) Desired upper limit for the price of the car (in dollars)

Your function should return a vector of car IDs that meet the specified criteria.

Problem 7-B.6

Write a function, **Prob7_B_6**, that inputs two arrays, **A** and **B**, of the same dimensionality. Your function should update the two arrays, and then output the (possibly) changed arrays **A** and **B**. The rules for updates are the following:

a) If an element in **A** is smaller than the corresponding element in **B**, set element in **A** to zero.

b) If element in **A** is greater than the corresponding element in **B**, set element in **B** to zero.

c) If element in **A** is equal to the corresponding element in **B**, set element in **A** to zero and set the element in **B** to zero.

Problem 7-B.7

Gillian, an agricultural engineer, is developing a new meal ration for emus. Gillian's goal is to create a mixture of corn, soybean meal, and cottonseed that will meet the nutritional needs of emus. The new ration will be sold in bags containing **proteinKG** kg of protein, **fatKG** kg of fat, and **fiberKG** kg of fiber. (Gillian wants to experiment with the values for each nutrient, so she is thinking of them as variables.)

The table below gives the nutrient content of each type of available raw material.

	Corn	Soybean Meal	Cottonseed
% protein	24%	42%	19%
% fat	41%	20%	21%
% fiber	31%	19%	11%

Write a function, **Prob7_B_7**, that will assist Gillian. Your function should input the value set for **proteinKG**, for **fatKG**, and for **fiberKG**. Your function should output (a) a vector containing the amount of corn to put in each bag, the amount of soybean meal to put in each bag, the amount of cottonseed to put in each bag, and (b) the total amount of feed in a single bag. The data in the table above should be represented internally in your function in an array.

Test your function with inputs from **Prob7_B_7(21,25,16)**. The total amount of feed you should compute for one bag of emu ration with these inputs should be 96 kg.

Once you complete your function for Gillian and have tested it, try the inputs of **Prob7_B_7(21,31,16)**. How can you tell from the result that this input set is not realizable as a bag of emu feed using the raw materials based on the results of your function?

Problem 7-B.8

Steel is often produced by blending carbon with other steel alloys that contain nickel, manganese, and chromium. The percentage of each of these metals in a blend affects the properties of the resultant steel. For instance, stainless steel is defined as a ferrous alloy (an alloy containing iron) with a minimum of 10.5 percent chromium content.

Suppose you are working for a steel blending company. Your company produces steel by combining three different stock alloys that each contain percentages of nickel, manganese, chromium, and carbon. The percentages for each of the three stock alloys your company uses, along with the cost per ton you have invested in each, are shown in the table below.

	Chromium (%)	Nickel (%)	Manganese (%)	Carbon (%)	Cost per Ton ($)
Alloy 1	25	7	5	2	115
Alloy 2	6	5	6	4	92
Alloy 3	11	6	7	5	129

You receive an order for 70 tons of stainless steel that is composed of 40% Alloy 1, 30% Alloy 2, and 30% Alloy 3. Does this alloy composition produce stainless steel? What are the overall chromium, nickel, manganese, and carbon percentages in the blend you will produce? What is the total cost of this production?

Instead of doing this problem from scratch each time you get a new order, your boss has asked you to write a function, **Prob7_B_8**, that inputs (a) the total number of tons in an order, (b) a vector containing the percentages of Alloy1, Alloy 2, and Alloy 3 specified in the order, and (c) a vector containing the cost you have in each ton for Alloy 1, Alloy 2, and Alloy 3. Your function should represent internally the percentages of chromium, nickel, and manganese for all of your stock alloys in an array.

Your function should output the following:

- A vector containing the computed number of tons of each of your three stock alloys to use to meet the order

- The cost you will have in raw materials to meet the order
- An answer to the question: Does the blend qualify as stainless steel? (return a logic value)

Point to think about: Why was it a good idea to utilize the cost of your base alloys as inputs in your function, while on the other hand the percentage content makeup of each stock material was represented internally in your function?

Problem 7-B.9

A furniture manufacturer makes dining tables, desks, coffee tables, and end tables. The material and labor requirement for each type of production is given in the table below:

	Dining Table	Desk	Coffee Table	End Table
Wood (sq ft)	145	130	95	55
Labor (hro)	12	9	7	5
Selling Price ($)	750	520	340	177

The current cost for wood is $2.25 per sq. ft.; the current cost for labor per hour is $8.50.

Your supervisor has asked you to write a function, **Prob7_B_9**, to process new orders. The orders will come in as quantities of dining tables, desks, coffee tables, and end tables. Your function should deal with these four numbers as a single input vector and as its first input argument. The second and third inputs to your function should be the current cost for wood and the current cost for labor.

Your function should compute and output (a) the total profit that you company will make on the order, (b) the total amount of wood used in the order, and (c) the total number of hours required to complete the order.

Problem 7-B.10

Your friend, Katherine, a stock broker, asks for your help in analyzing stock prices. She tells you that she keeps daily records for a number of stocks. Her strategy is to buy stocks whenever the stock price is lower than its average over the last three days and to sell a given stock if its price is higher than the three-day average.[6]

Katherine sends you a sample record of stock prices as shown in the table below. The first column represents the stock identifier and the remaining columns are stock prices in successive days.

Stock Number	Price ($) April 21st	Price ($) April 22nd	Price ($) April 23rd	Price ($) April 24th
001	14.00	15.25	13.50	15.50
002	21.00	18.50	19.25	19.00
003	7.00	7.75	7.25	7.50
004	32.50	31.25	30.00	30.25

Write a function, **Prob7_B_10**, that inputs the stock prices in the table as an array. Your function should compute and output two vectors:

- The identifiers for stocks that should be bought on April 25
- The identifiers for stocks that should be sold on April 25

Each day Katherine wants to play the market with her strategy, she is going to have to retype the array with current stock prices. This would get old, fast. What would you like to be able to do to prevent retyping? How could you implement your idea by extending your function? (For one possible option, see **save** and **load** in MATLAB Help.)

---

6. The first thing you should tell Katherine is that she is likely to lose her shirt with this strategy!

Chapter 8

More Flexibility: Introduction to Conditional and Iterative Programming

Your progress in learning to use MATLAB for technical problem solving up to this point has focused largely on gaining proficiency in representing and solving technical problems using MATLAB numerical scalars, numerical vectors, and numerical arrays. As you saw first in Chapter 5 on numerical vectors, then in Chapter 7 on numerical arrays, MATLAB gives you the power to perform a combination of simple operations by executing a single MATLAB command.

For example, if you have a set of angles $\left[-\frac{\pi}{2}, -\frac{\pi}{3}, -\frac{\pi}{6}, 0, \frac{\pi}{6}, \frac{\pi}{3}, \frac{\pi}{2} \right]$ (in radians) and want to compute the hyperbolic sine and the hyperbolic cosine for each of the angles, the following MATLAB code will accomplish your task:

```
>> angleVector = [-pi/2, -pi/3, -pi/6, 0, pi/6, pi/3, pi/2];
>> angle_hSin_hCos = ...
        [angleVector', sinh(angleVector)', cosh(angleVector)']
angle_hSin_hCos =
        -1.57           -2.30           2.51
        -1.05           -1.25           1.60
        -0.52           -0.55           1.14
            0               0           1.00
         0.52            0.55           1.14
         1.05            1.25           1.60
         1.57            2.30           2.51
```

If you were to use a calculator to compute the entries in the array **angle_hSin_hCos_Array**, you would need to do 14 separate computations, one for each entry in the second and third columns of **angle_hSin_hCos_Array**. With MATLAB, you can accomplish all 14 computations with one command. MATLAB enables you to treat all elements of a vector at one time and perform the same operation on all elements.

Suppose you were asked to compute a table of horizontal travel distances for a projectile given initial speed (in m/s) and firing angle (in degrees) pairs as follows:

$[(10, 45), (20, 60), (30, 75), (40, 85)]$; and using the equation $D = \dfrac{2v_0^2 \sin(\phi)\cos(\phi)}{g}$

where D is the horizontal travel distance, v_0 is the initial speed, and ϕ is the firing angle. MATLAB makes performing a number of simple computations easy:

```
>> initSpeedVector = [10, 20, 30, 40];
>> firingAngleVector = [45, 60, 75, 85];
>> fAngleRad = firingAngleVector * (pi/180);
>> hDist = 2 * initSpeedVector.^2 .* sin(fAngleRad) .* cos(fAngleRad);
>> speed_angle_hDist = [initSpeedVector', firingAngleVector', hDist']
speed_angle_hDist =
        10.00          45.00          100.00
        20.00          60.00          346.41
        30.00          75.00          450.00
        40.00          85.00          277.84
```

Both of the examples above show the power of cell-by-cell computations using MATLAB. The catch is that cell-by-cell computations are applicable only when the corresponding elements of vectors or arrays are the starting point(s) of a computation. For example, in the first example above, **angleVector** contains a set of angles we want to use to compute hyperbolic sine and hyperbolic cosine. For each angle, we compute a corresponding value for the hyperbolic sine, and we compute a corresponding value for the hyperbolic cosine. Construction of the array **angle_hSin_hCos_Array** was easy because all we had to arrange was an array in which each row listed the following:

<value of angle>, <corresponding value of hyperbolic sine>, <corresponding value of hyperbolic cosine>

In the second example, we used two input vectors (**initSpeedVector** and **firingAngleVector**) containing corresponding scalar values and used the two input values to compute the corresponding scalar output for one element of **hDist**.

One row in the answer array for both of the above examples represent one instance of the problem.[1] In the projectile problem, for example, the second row represents a solution for the scalar problem: Given a firing angle of 20 degrees and a muzzle velocity of 60 meters/second, we obtain a horizontal flight distance of 346.41 meters. Vector/array solutions done cell by cell in MATLAB are paths for doing many problem *instances* with one command.

---

1. We could organize these solutions so columns were used to represent one problem instance. A useful exercise would be to go back and reformulate both problems so that one column represents one problem instance.

One characteristic of these examples makes them prime targets for solving by vector/array manipulations using cell-by-cell operations. In both examples, the computations you do in one row (that is, in one problem instance) are independent of computations in any other row. In fact, cell-by-cell operations fit problems with this characteristic beautifully because the definition of cell-by-cell operations demands this independence. This is an important point.

Unfortunately, not all problems are of this nature. Not all large problems can be understood as a problem where each row of a problem solution represents a problem instance, and where each problem instance does not depend on any other instance. Next, we will examine a problem in which the solution requires more flexible programming capabilities than you have learned so far.

The goal in this chapter is to extend your ability to handle more complex problems by using MATLAB loops and conditional structures. Let us examine a problem that will set the stage; we will use this example throughout much of this chapter.

8-1   A Focus Problem

Before diving into the MATLAB topics of this chapter, we'll begin by setting up a problem. We will use this sample problem multiple times through this chapter, so be sure you understand the set up as described below.

Assume that you are the Inventory Manager for the Big League Bats Company. Each week, your factory produces a number of baseball bats that are moved to your company warehouse. All bats sold come from the inventory on hand, not from the factory. Suppose Big League Bats has a fiscal year that starts on June 1. You identify each week by its number starting this year from June 1, and as luck would have it, June 1 is a Monday. So, Week 1 starts on Monday, June 1. Week 2 starts on Monday, June 8, etc.

A computational model describing the week-by-week changing inventory as a function of the week number is as follows:

$$Inventory(k+1) = Inventory(k) + Production(k) - Sales(k) \qquad \textbf{(EQ 8-1)}$$

The terms on the right-hand side of Equation 8-1 refer to a specified week, k. The left-hand term refers to the next week after k, that is, $k+1$.[2] Equation 8-1 is saying that the inventory we start with *next* week will be what we start with this week plus how many bats we make this week minus how many bats we sell this week. Keep in mind we are using the word inventory to mean starting inventory. *Inventory(k)* means the starting inventory for week k. *Inventory*($k+1$) means the starting inventory for week $k+1$.

Your goal as Inventory Manager is to have enough bats to meet demand, but to keep inventory as low as possible. (Why do you have that goal?)

You do not control sales, and no one can predict what sales will be with 100 percent accuracy. But the Sales Manager projects weekly sales based on past years. For the first five weeks of the new fiscal year, the Sales Manager predicts sales will be good, largely because of high demand for bats as summer is under way. Specifically, he predicts sales for the first five weeks of the new fiscal year as shown in the second column of Table 8-1.

---

2. Equation 8-1 is an example of a relation that a mathematician terms a "recurrence relation." You may have already studied recurrence relations, but under another name: mathematical induction. Check out "recurrence relation" using Google.

Table 8-1: Projected Sales and Projected Factory Output for Big League Bats - Data of the Problem

Week Number	Projected Number of Bat Sales	Projected Number of Bats Manufactured	Projected Ending Inventory for the Week
1	50,000	45,000	?
2	55,000	50,000	?
3	60,000	55,000	?
4	70,000	60,000	?
5	45,000	65,000	?

As with sales, you do not control production either. The Factory Manager does. When you inquire, you learn that the manager uses a simple means of projecting how many bats to produce in a specific week by organizing the factory to produce a number of bats equal to the sale of bats in the last week. Except for Week 1, you can see that the data in the third column of Table 8-1 repeats the sales projection quantity, but offset one row down. You can find the projected manufacture rate for Week 1 by getting sales data from the last week; that value (shown in Column 2 of Table 8-1) is 45,000 bats.

To meet your goal (to cover demand, but to keep inventory to a minimum), you refer to the data in Table 8-1. Look back at Equation 8-1. It tells you how to compute the inventory for the start of each week given the inventory for the start of the preceding week, the sales for the preceding week, and the production for the preceding week. For example, applying Equation 8-1 to find the starting inventory for Week 3, you take the starting inventory for Week 2 minus the number of bats sold in Week 2 plus the number of bats made in Week 2.

What else do you need to know to project how the inventory will fluctuate given the data in Table 8-1? What do you control as Inventory Manager?

The only data needed is the starting inventory for Week 1. You are the Inventory Manager, so you can set this value. Suppose that the starting inventory for Week 1 is 50,000 bats. Then the starting inventory for Week for Week 2, is as follows:

$$Inventory(2) = Inventory(1) + Production(1) - Sales(1)$$
$$= 50000 + 45000 - 50000 \qquad \text{(EQ 8-2)}$$
$$= 45000$$

Repeated application of Equation 8-1 (for Week 2, Week 3, etc.) results in finding the

Table 8-2: Projected Sales and Projected Factory Output for Big League Bats (Solution of the Inventory Problem) (with starting inventory for Week 1 is 50,000 bats)

Week Number	Projected Number of Bat Sales	Projected Number of Bats Manufactured	Projected Ending Inventory for the Week
1	50,000	45,000	45,000
2	55,000	50,000	40,000
3	60,000	55,000	35,000
4	70,000	60,000	25,000
5	45,000	65,000	45,000

starting inventory for all weeks as shown in Table 8-2. Work out the solution for each week's starting inventory with pencil and paper and be sure your computations agree with the results shown in Table 8-2.

Look at what you have done with pencil and paper to solve the inventory problem for Big League Bats. Each row of Table 8-2 is a problem instance in that each row represents a problem and a solution for one week of the inventory problem you, as Inventory Manager, have to deal with. But the rows (the individual solutions for each week) depend on your solving the prior week's problem because your solution uses the result from the preceding week.

You might wonder why you should bother with MATLAB at all if this problem is so simple to solve with paper and pencil. The answer is that a real inventory control problem in manufacturing is typically more complex than solving for only five weeks; you might need to do an entire year's projection. And the method used to determine the level of production would typically be more complex than what we have used.

The inventory problem above is, in general, a problem that you cannot solve in a straightforward manner with cell-by-cell operations. The purpose of this chapter is to give you the MATLAB programming tools you need to solve such problems. But, in order to understand the big picture, we need to take a side journey before talking about the MATLAB mechanisms for solving problems like the Big League Bats problem. In the next section, we will set the stage by extending the one type of program control you have learned so far to include two new types of program control.

8-2   General Picture, Program Flow

So far, our development of MATLAB user-defined functions has been limited to building functions that are examples of straight line code. The general picture illustrating straight line code is shown in Figure 8-1. Each one of the boxes represents a single MATLAB command as it would be typed in to the Command Window. Those could be simple commands with operators:

```
z = x + y
```

They could be commands that include built-in functions:

```
shouldBe1 = sin(x)^2 + cos(x)^2
```

They could be commands that include user-defined functions:

```
oscAns = sin(x) * myFunction(y + z^2)
```

Or, they could be any command that is a valid MATLAB expression. In each case, the general story is the same. Each of the commands in the boxes in Figure 8-1 stands in for a MATLAB command.

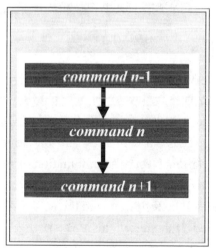

Figure 8-1: Flow of Straight Line Control

In this chapter, you will learn two new control structures for MATLAB code: conditional structures and looping structures.[3]

Look at the general picture for conditional control shown in Figure 8-2. The top and bottom boxes (*command n*, and *command m*) can be thought of as MATLAB

---

3. Looping structures are also called iterative structures.

commands. The grey box in Figure 8-2 is a conditional construct. The best way to think of the *entire* grey box is that it is simply as another command that is executable in MATLAB. After *command n* is executed, then the entire background box is executed and *command m* is executed. So, in one sense, there is nothing new here. But, in another sense, the contents of the grey box give MATLAB alternative paths of execution.

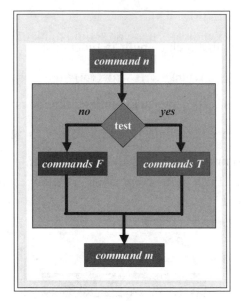

Figure 8-2: Flow of Conditional Control

The grey box in Figure 8-2 starts out with a test (the diamond-shaped box). This test will involve a MATLAB relational test or a combination of relational tests. Depending on the result of the test, one of the alternative paths is chosen. If the test yields **TRUE**, then the path marked **YES** will be followed, and the MATLAB command(s) in the box on the right will be executed. If the test yields **FALSE**, then the path marked **NO** will be followed, and the MATLAB command(s) in the box on the left will be executed.

Emphasizing a point made above, you should think of the entire contents of the grey box in Figure 8-2 as *one conceptual MATLAB command*. Program flow enters the grey box at one point (from the *command n* box) and leaves from one point (going to the *command m* box). This view of the conditional construct is rooted in your gaining the ability to modularize your MATLAB programs into blocks with a single entry and single exit so you can think of your code at a high level as straight line code, where elements of the straight line code are each MATLAB commands or conditional constructs.

The good news is ...Conceptually, that is all there is to a conditional construct. The less good news is ... You have more work to do to learn how to implement the concept shown in Figure 8-2. The way to turn the conceptual picture of conditional constructs into working MATLAB programs is the subject of Section 8-4.

You need to learn one additional type of control flow construct: iterative control.[4] Figure 8-3 shows the concept of iterative program control flow. As in the case of conditional program control flow, we will use an enclosing background box. As before, the background box indicates that you should think of the iteration construct as a MATLAB block with one entrance point and one exit point. The difference between conditional and iterative constructs starts with their purposes.

For conditional constructs, the purpose is to make a decision about which alternative code commands to execute. For iterative code constructs, the purpose is to reuse the same code (potentially) more than once.

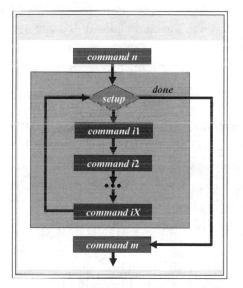

Figure 8-3: Flow of Iterative Control

The diamond in Figure 8-3 contains commands to set up variables to be used in the enclosed-box commands which are part of the iterative construct. Once the necessary

---

4. We will not discuss recursive control. In a nutshell, recursive user-defined functions are those that can call themselves. Recursive functions form an entire branch of mathematics and give an alternative to iterative constructs. In some cases, recursive user-define functions in MATLAB are a natural way to develop solutions for some problems.

variables are set, commands in the enclosed boxes are executed one at a time until the end of the construct is reached (*command iX* in Figure 8-3). Control goes back to the top of the loop, the variables needed to run the loop are reset, and the commands in the loop are executed again. The diamond setup is also used to determine if the loop is done. If it is, then control moves on to the single exit point and to the next command outside the iterative construct.

Think back to the problem of inventory control at Big League Bats (Section 8-1), and in particular to the data table we built to represent the problem (Table 8-2). Remember that each row of the data table is a statement of data for one instance of the problem we want to solve. Note that we want to solve this same problem multiple times, once for each data table row. The setup diamond in Figure 8-3 needs to set the beginning inventory for each week (which is the same as the ending inventory from the previous week). The application of Equation 8-1 to the beginning inventory fixed by the setup and the data from the proper row of the data table for Big League Bats gives the result we want: a computed value for the starting inventory for the next row of the data table. Applying Equation 8-1 is done conceptually by the enclosed background boxes in Figure 8-3. Once a pass through the commands of the enclosed loop is completed, the program control goes back to the top of the loop, the setup readies the variables for computing the next pass through the loop, and we proceed through the same code again. In computer talk, we *iterate* the code.

As was the case for conditional constructs, you have to learn the syntax for iterative constructs in MATLAB. If you understand the concept of the iterative construct as shown in Figure 8-3, then you will understand the concept behind the last paragraph in the solution of the inventory problem for Big League Bats. In Section 8-3, you will find the specifics for building iterative constructs in MATLAB. You will see that there are two major types of iterative blocks: FOR loops and WHILE loops.

Keep in mind the idea of blocks (or modules) being the building blocks for MATLAB code. Be sure to note that the background block in Figure 8-3 that covers the entire iterative construct has one entry point and one exit point. Now that we have discussed the general nature of the iterative construct, it is possible to think of your code, at a high level, as straight line code, where elements of the straight line code are simple MATLAB commands, conditional constructs, or iterative constructs.

1. There are three major types of program control: straight line control, conditional control, and iterative control.

2. Programming constructs for conditional control and iterative control should be considered "modules," meaning there is one point of entrance into the construct and one point of exit. Thinking of programming constructs in this way increases the modularity of code, and that, in turn, increases your ability to understand, extend, and debug code.

3. Straight line code executes in the order it is written in a program.

4. Conditional code executes one alternative of a number of possibilities, selecting the alternative to run based on a relational/logical test of program variables.

5. Iterative code executes the same block of code a number of times. There are two main types of iterative loops: FOR loops and WHILE loops.

Synopsis for Section 8-2

8-3   Iterative Program Flow: For LOOP

The task of this section is to start with the general picture for iterative control shown in Figure 8-3 and specify the syntax that MATLAB requires of you when you want to write an iterative loop. In this section, we will spend most of our time examining the **FOR** loop construct, but we will look at the **WHILE** loop construct too. Both will give you useful additions to your set of MATLAB capabilities.

A common use for a **FOR** loop is to "step through" the elements of a vector and perform a computation with each element in turn. Iterating through the elements of a vector will be the first concept we will discuss. Most computer languages give you the ability to iterate through elements of a vector though most do not enable you to deal with this iteration as directly as MATLAB does. This is the subject of Section 8-3.2.

In Section 8-3.3, we will focus on **FOR** iteration through the columns of an array. Few computer languages other than MATLAB give you a direct operation for stepping through columns of an array one by one.

In Section 8-3.4, you will learn how to nest **FOR** loops, that is, how to put one loop inside another with the purpose of gaining complete control of accessing elements of an array.

Finally, in Section 8-3.5, we will tie together an understanding of iterative control in MATLAB by using **FOR** loop programming to solve the inventory problem for Big League Bats, the ongoing problem introduced in Section 8-1. Let's start in Section 8-3.1 with the general form of **FOR** loops in MATLAB to give you a framework on which to build.

8-3.1   General Form of a FOR Loop in MATLAB

The general form of a **FOR loop** in MATLAB is shown in Figure 8-4. The parts of Figure 8-4 shown in italics are what you must fill in with specifics. The two MATLAB keywords (**for** and **end**) and the equals (**=**) sign must be part of a valid **FOR** loop.

You could describe a **FOR loop** this way:

> *A block of MATLAB code starting with the MATLAB keyword* **for** *and ending with the MATLAB keyword* **end** *and enclosing one or more MATLAB commands. The* **loopVar** *sequentially takes on values in* **loopVectorOrArray** *as the sequence of commands enclosed in the*

```
for loopVar = loopVectorOrArray
    Command_1
    Command_2
        ⋮
    Command_N
end
```

Figure 8-4: General Form of a FOR Construct in MATLAB

FOR *loop are run again and again. The loop completes when all* *elements in* **loopVectorOrArray** *have been used.*

The code line that starts the loop looks much like an assignment statement following the keyword **for**. In fact, it is an assignment statement but not like any you have seen before. A normal assignment statement (like **x = 22**) sets the result of evaluating some MATLAB expression on the right of the equals sign to be the value of some MATLAB variable on the left hand side of the equals sign.

But the part of a **FOR** loop that looks like an assignment statement sets the value of a MATLAB variable (the **loopVar**) to successive values of the vector or array (**loopVectorOrArray**) on the right-hand side of the equals sign. If **loopVectorOrArray** is a vector,[5] then **loopVar** is set initially to the value of the first element of **loopVectorOrArray**, and the body of the loop (**command1**, **command2**, etc., **commandN**) is run using that setting for **loopVar**. Then **loopVar** is reset to the value of the second element in the vector. The whole process is repeated until all elements of **loopVectorOrArray** have been processed.

The purpose of an iteration is to run and rerun a block of commands. The **FOR** loop construct in MATLAB gives an organized way to run the same code block (the body of the **FOR** loop) repeatedly, each time setting the value of an identified variable (the **loopVar**) to an element of a specified vector. As we turn to specifics below, keep in mind the general structure of **FOR** loops as shown in Figure 8-4. By the end of this section on **FOR** loops, you should understand the variations of **FOR** in terms of the general picture presented in Figure 8-4.

---

5. We will show in Section 8-3.3 what happens if **loopVectorOrArray** is an array.

8-3.2　Iteration Over Elements of a Row Vector Using FOR

Suppose you want to compute the factorial of an integer N. The following MATLAB function performs the required computation[6]:

```
1 function theFactorial = practiceFactorial(N)
2 % computes the factorial of a non-zero, positive integer N
3 % INPUT: N - a non-zero, positive integer
4 % OUTPUT: theFactorial - the computed value of N!

5   % initialize theFactorial and the looping vector
6   theFactorial = 1;
7   loopingVector = 2:N;

8   % compute theFactorial using a FOR loop
9   for oneValue = loopingVector
10      theFactorial = theFactorial * oneValue;
11      disp([oneValue, theFactorial])
12  end
```
[7]

Lines 9-12, shown inside the box, form the loop conceptually depicted in the background box of Figure 8-3. Line 9 starts with the MATLAB keyword **for**, followed by what looks like an assignment statement. Line 10 is a standard MATLAB command, as is Line 11. Line 12 is the single MATLAB keyword **end**.

Here are the first of a number of rules we will present for **FOR** loops:

- *Rule* 1 for **FOR** loops: **FOR** loops start with the keyword **for** followed by what looks like an assignment statement.

- *Rule* 2 for **FOR** loops: **FOR** loops *must* end with the MATLAB keyword **end**.

So, **FOR** loops start with **for** and end with a line saying **end**. What comes in between are the MATLAB commands to be repeated each time the loop is iterated.

Go back to Line 9. The part of Line 9 that looks like an assignment operator is not a standard type of assignment command. Instead, each time through the loop, the "next" value of the vector on the right side of the "pseudo-assignment" will be set to the variable on the left side. The variable on the left of the operator in a **FOR** line is called the *loop variable*.

---

6. MATLAB includes a built-in function called **factorial**. We are building our own here for exposition. However, we do not use the function name factorial in the one we are building.

7. Look up the MATLAB built-in function **disp** to understand how it is used in this line.

The first time through the loop (Lines 9-11) the loop variable, **oneValue**, will be set to the first element of vector **loopingVector**. The second time through the loop, **oneValue** will be set to the second element of vector **loopingVector**. This will continue until we have gone through all the elements of **loopingVector**.

Implement **practiceFactorial** on your computer. Line 11 displays interim values each time the loop is executed. Once you have the implementation done, execute the following command:

```
>> theAnswer = practiceFactorial(4)
```

You should see the following in your command window:

```
        2       2
        3       6
        4      24
  theAnswer =
        24
```

Be certain you understand how this output is generated.

Note this easily missed, important point about iterative loops: Vectors, like **loopingVector** above, that are used to hold the elements used in successive iterations of a loop must be row vectors. Consider the following code fragment containing a **FOR** loop:

```
for i = 1:3
    i
end
```

When this loop is executed, all that is done is to display the value of the loop variable:

```
i =
    1
i =
    2
i =
    3
```

In contrast, consider this loop:

```
for i = [1; 2; 3]
     i
end
```

It produces the following:

```
i =
     1
     2
     3
```

In the first case, a row vector was used in the **for** statement. The result was that the loop executed three times, and the value of **i** was 1 the first time through the loop, 2 the second time, and 3 the third time. In the second case, a column vector was used in the **for** statement. The loop executed only once, and the value of **i** that one time was the entire column vector.

- *Rule* 3 for **FOR** loops: When the desired effect of using a **FOR** loop is to iterate over the elements of a vector, one element at a time, a row vector must be used.

As you can see, there is nothing magic about **FOR** loops. They can be used to methodically step through the elements of a vector, use each element in turn as the value for the loop variable, and execute the same set of MATLAB commands within the loop.

8-3.3   Iteration Over Columns of an Array Using FOR

In the last section, you learned how the MATLAB **FOR** loop can be used to repeat the same code block, such that each iteration uses a new value of the loop variable in the **FOR** line. In this section, you need to make only one extension: how to use **FOR** loops to iterate through the columns of an array.

Suppose you have the array **A** as follows: $\begin{bmatrix} 1 & 2 & 3 \\ 10 & 20 & 30 \end{bmatrix}$ and you want to multiply the values in each column and add the product results for each column to produce a final result. The following function would accomplish that task for the two-row array **A**:[8]

---

8. **sumColumnProducts** could be implemented using the built-in function **cumprod**—and in such a way that it would allow any size **A** as input—without using a **FOR** loop. We are using this example to show iteration over the columns of an array. See if you can develop the more general solution without **FOR** loops using **cumprod**.

```
function sumColumnProducts = computeColSum(A)
% computes products for each column, and returns
% sum over all columns
%    valid only for 2-row arrays
% INPUT:  A - an array with any number of columns and two rows
% OUTPUT: sumColumnProducts - a scalar containing resultant sum
% SAMPLE INPUT/CALL/OUTPUT
%  aSampleInput = [1, 2, 3;  10, 20, 30];
%  aSampleOutput = computeColSum(aSampleInput)
%  aSampleOutput => 140

    sumColumnProducts = 0;
    for oneCol = A
        sumColumnProducts = sumColumnProducts + ...
        oneCol(1) * oneCol(2);
    end
```

When run, this function will result in the following display in your command
window…

```
>> aSampleOutput = computeColSum(aSampleInput)
aSampleOutput =
    140
```

In **sumColumnProducts**, we accessed the values in the loop variable *vector* in the
body of the loop the same way we would access any vector. If you think back to the
last section, the way that MATLAB sets the loop variable to successive columns in an
array explains why you must use a row vector if you want to iterate over a vector.

- *Rule* 4 for **FOR** loops: When an array is used on the right-hand
  side of the pseudo-assignment statement in a **FOR** construct, the
  loop variable is successively set to the columns of the array.

8-3.4   Nested FOR Loops in MATLAB

In the previous section, you learned how to iterate over columns in an array. Many
times, the technique shown for this type of iteration will be enough to let you perform
computations using all array elements. Sometimes, this is not enough control.

Suppose you are given an array **A**. You are asked to create a new array **X** such that all
the elements in the first row and the first column of **X** are the same as the
corresponding elements in **A**, while all other elements in **X** are computed by adding to
A the corresponding element to the left, the corresponding element diagonally to the
upper left, and the corresponding element above. For the cell in the third row, fourth
column of **X** (as shown in Figure 8-5):

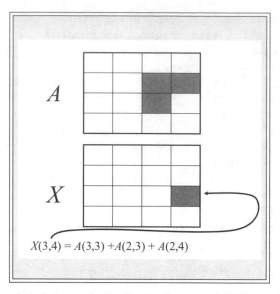

Figure 8-5: Schematic for Producing Array X Given Array A

$$X(3,4) = A(3,3) + A(2,3) + A(2,4)$$

The following function will create array **X** as defined above:

```
1   function resA = createSpecialArray(A)
2   % construct an array in which each elements for rowNum > 1,
3   % colNum > 1 are calculated by adding cell to left, cell above, and
4   % cell to upper left diagnonal from original input array
5   % INPUT: A - array of arbatrary size, numRows>2, numCols>2
6   % OUTPUT: resA - the modified array per above
7
8       resA = A;
9       [nRows, nCols] = size(A);
10      for i = 2:nRows
11          for j = 2:nCols
12              resA(i,j) = A(i, j-1) + A(i-1,j-1) + A(i-1, j);
13          end
14      end
```

The lines doing the heavy lifting of **createSpecialArray** are marked as lines 8-14, which is where we will concentrate:

- **Line 8** sets the output variable to be equal to the input variable. Remember we want to leave Row 1 and Column 1 of the input unchanged in the output. Line 1 assures that we start out by meeting that requirement. The rest of the code will modify all but the first row and first column of the output per the problem specifications.

- **Line 9** sets variables that hold the number of columns and the number of rows in the input because we want the function to work regardless of **A**'s size.

- **Lines 10-14** are two loops, one enclosed by the other, i.e., nested loops. (Note the boxes.)

    - **Line 10**, the block of **Lines 11 -13**, and **Line 14** taken together form the outer loop. The loop variable for the outer loop, **i**, will count from 2 up through the number of rows in the input array.

    - **Lines 11-13** form the inner loop. The loop variable for the inner loop, **j**, will count from 2 up through the number of columns in the input array.

To understand the operation of these nested loops, start by examining the code below:

```
A = [1 2 3 4; 5 6 7 8; 9 10 11 12];

[nRows, nCols] = size(A);

for i = 2:nRows
    disp( sprintf('i=%g',i) );    % prints value of i on one line
    for j = 2:nCols
        disp( sprintf('  j=%g',j) );% prints value of j on one line
    end
end
```

If you do not remember how to use **sprintf**, use MATLAB Help to refresh your memory.

Implement the code above in a script file on your computer and name it **nestedLoopsScriptCounter**. Call the script. You should get the following result:

```
>> nestedLoopsScriptCounter
i=2
    j=2
    j=3
    j=4
i=3
    j=2
    j=3
    j=4
```

The operation of the nested loops could be described this way: First, **i** takes the value 2. The **j**-loop is run giving **j** a value of 2 its first time through, 3 the next time, and finally, **j** has the value 4. That completes the **j**-loop, so control goes back to the outer **i**-loop and the second run is started with **i** taking a value 3. Again the same **j**-loop is run.

Understanding MATLAB loops, especially nested MATLAB loops, is strongly dependent on your ability to understand computers. That is, when you see a loop (a nested loop) you need to start stepping through it in your mind as MATLAB would during execution. If you develop that skill, any mystery or confusion about loops is likely to vanish.

The fragment we put into the script file **nestedLoopsScriptCounter** was similar to the working part of the function **createSpecialArray**. Though the code in **nestedLoopsScriptCounter** was aimed at cycling through the indices of an array and printing values of those indices, the code in **createSpecialArray** used the values of the indices to create the desired output.

8-3.5   Solving the Big League Bats Problem: Version 1

We can now return to the inventory problem for Big League Bats (Section 8-1). Think of solving this problem as concluding this section on building MATLAB **FOR** loops and as a reality check for how well you understand iteration.

Suppose you are asked to write a user-defined function that will solve the Big League Bats problem. The explicit problem parameters are as follows:

- Inputs are (a) a scalar for the starting inventory for Week 1 and (b) an array whose rows are week-by-week values for "Projected Number of Bat Sales" (column 1 of Table 8-2) and "Projected Number of Bats Manufactured" (column 2 of Table 8-2).

- Output is a vector containing "Projected Ending Inventory" (column 3 of Table 8-2).

- The function will create a well-labeled plot of "Projected Ending Inventory" versus week.

The following function definition meets the set requirements:

```
1   function projEndInventory = projectBigLeagueBats(startInv, blbData)
2   % projects ending weekly inventory for BLB
3   % INPUTS: startInv - scalar, the starting inventory for Week 1
4   %         blbData - array, rows are weekly data for BLB, where each row
5   %         is [batSales, batsManufactured]
6   % OUTPUT: projEndInventory - vector of num of bats at end each week
7   % SAMPLE SET UP AND CALL
8   %      week1StartInv = 50000;
9   %      blbData = 10^3 * [50,45; 55,50; 60,55; 70,60; 45,65];
10  %      endInv = projectBigLeagueBats(week1StartInv, blbData);
11
12     projEndInventory = [];       % initialize the output variable
13     startInvThisWeek = startInv;% initialize a variable to hold
14                                 % a scalar variable that holds the
15                                 % starting inventory for each week
16  % NOTICE IN THE FIRST LINE OF THE LOOP (19), blbData is transposed.
17  % BE SURE YOU UNDERSTAND WHY!
18
19  for oneWeekSet = blbData'
20     batSalesThisWeek = oneWeekSet(1);
21     batsManuThisWeek = oneWeekSet(2);
22     endInvThisWeek = startInvThisWeek+batsManuThisWeek-batSalesThisWeek;
23     projEndInventory = [projEndInventory, endInvThisWeek];
24
25   % update startInv to be the starting inventory for NEXT week
26     startInvThisWeek = endInvThisWeek;
27  end
28
29  % create the required data plot
30     createBigLeagueBatsPlot(startInv, projEndInventory)
```

The code for **createBigLeagueBatsPlot** (called in line 30) will be shown
below. It is good programming practice in MATLAB, or any computer language, to
hold the total number of lines of code in a user-defined function to a reasonable size. It
is a good rule of thumb to subdivide your code if you find a user-defined function
extending past fifty lines of code.

Calling **projEndInventory** with the sample variable values written into the function (lines 8-9) gives the result below:

```
>> startInv = 50000;
>> bigLeagueBatsData = 10^3 * [50,45; 55,50; 60,55; 70,60; 45,65];
>> finalInv = projectBigLeagueBats(startInv, bigLeagueBatsData)
finalInv =
        45000        40000        35000        25000        45000
```

This agrees with the results of the hand computations shown in Table 8-2.

Implement **projEndInventory** on your computer, and comment out line 30 for now. Be sure you understand each line of code that you build: It will do you no good to copy the code into your MATLAB editor.

In discussing **projEndInventory**, first, be certain you understand the purpose of line 13. (Hint: Look at line 30.) **startInv** holds the value for the Week 1 beginning inventory of bats and is an input variable to the function. **startInvThisWeek** is a variable that will hold the value for the number of bats that we start out with each week. For Week 1, the two variables have the same value, insured by the command of line 13. Line 30 is a call that uses the Week 1 starting inventory; hence, we want to make sure we do not overlay the value originally handed to the function.

Second, be sure you understand Lines 19-22. In Line 19, a **FOR** loop is set up. The loop variable is **oneWeekSet**. **oneWeekSet** cycles through columns of the input array **blbData**, and rather than use **blbData** directly, we use its transpose. The transpose is taken because the original order of **oneWeekSet** is row order, meaning that data instances are organized so that each instance is one row in the array. (Look back to Table 8-2 to confirm this.) **FOR** loops in MATLAB cycle through columns of an array and, hence, need to be given an array that is organized in column order, which means that data instances are organized so that each data instance is one column in the array. To get from row order organization to column order organization, the MATLAB transpose operator is applied. Lines 20-21 are to pull data out of the loop variable vector **oneWeekSet** so the code is more easily understood.

Third, line 22 is a straightforward expression: For a given week, the ending inventory is the starting inventory for the week, plus the number of bats manufactured that week, less the number of bats sold that week.

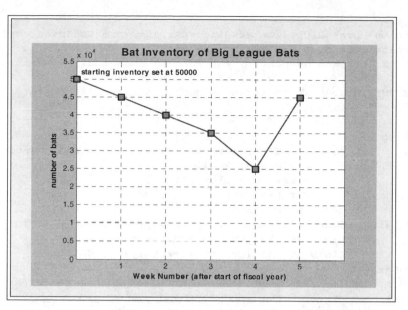

Figure 8-6: Data Plot for Projected Inventory, Big League Bats

Fourth, line 23 resets the growing vector that holds the output result—the values for ending inventory, week-by-week—that **projectBigLeagueBats** will return. The reset adds a new item for the current week to the end of the vector of computed values.

Fifth, line 26 is key. Line 26 is what makes this whole **FOR** loop work. In line 26, the starting inventory for the next week is set equal to the ending inventory for this week. Why is this done for next week? Because line 26 is at the end of the loop, and line 26 is setting things up for the next pass through the loop. This strategy is common in building **FOR** loops. We need iterative structures for problems like the Big League Bats inventory problem, because we need the result of one data instance to solve the succeeding data instance. For this inventory problem, we need the ending inventory for one week to be used as the starting inventory for the next week. That identification is what is done in line 26.

Now we go on to deal with creating the data plot as a required part of this problem. Code for **createBigLeagueBatsPlot** is shown below:

```
1   function createBigLeagueBatsPlot(startInv, projEndInventory)
2   % create data plot for Big League Bats
3   % INPUT: projEndInventory - the week-by-week ending inventory
4   %                           of Big Leauge Bats
5   % OUTPUT: no output variables - but data graph is generated
6
7
8     figure
9     plot(0:length(projEndInventory), [startInv, projEndInventory], '--rs',.
10                    'LineWidth',2,...
11                    'MarkerEdgeColor','k',...
12                    'MarkerFaceColor','g',...
13                    'MarkerSize',10)
14    set(gca, 'xTick', 1:length(projEndInventory), ...
15             'XLim', [0, length(projEndInventory) + 1], ...
16             'YLim', [0, max(projEndInventory)+10000])
17    grid on
18
19    xlabel('Week Number (after start of fiscal year)', 'FontWeight', 'bold')
20    ylabel('number of bats', 'FontWeight', 'bold')
21    title('Bat Inventory of Big League Bats', ...
22              'FontWeight', 'bold', ...
23              'FontSize', 14)
24
25    annotationStr = sprintf('starting inventory set at %g', startInv);
26    text(0.10, startInv + 2000, annotationStr, ...
27              'FontWeight', 'bold')
```

Most of the code for **createBigLeagueBatsPlot** should be familiar to you. In the call to **plot** (lines 9-13), there are properties for lines that we have not directly introduced. (See MATLAB Help if you need to.) Lines 14-16 are calls to built-in function **set**. The purpose of **set** is to set variables associated with a current plot, called the *current axis*. The current graphical axis is found by calling another built-in function **gca** (for "get current axis"). Consult MATLAB Help for a full description.

Implement **createBigLeagueBatsPlot** on your computer. Go back to **projectBigLeagueBats** and uncomment line 30. Then, execute the following in the Command Window:

```
>> startInv = 50000;
>> bigLeagueBatsData = 10^3 * [50,45; 55,50; 60,55; 70,60; 45,65];
>> finalInv = projectBigLeagueBats(startInv, bigLeagueBatsData)
```

In addition to producing the output variable **finalInv**, you should produce a data plot as shown in Figure 8-6.

Let's return to the original purpose that was set in the inventory problem for Big League Bats. The purpose was to meet sales needs and to have the minimum number of bats in inventory to meet those needs. The problem was such that the only variable you (as Inventory Manager) controlled was the starting inventory for Week 1. Examine Figure 8-6. Did you set the starting inventory too high, too low, or about right? To what value should the original starting inventory be set to meet sales while keeping inventory minimal? Test your answer by using **projectBigLeagueBats**.

8-3.6   Synopsis of MATLAB FOR Loop

If you need to perform an array computation that involves data values from the same data instance, then you should use cell-by-cell operations. But suppose, instead, you need the results of the computation on the preceding data instance to set up a computation for one data instance. Then you must properly use **FOR**.

FOR loops are one of three major types of iterative structures in MATLAB. Another type is **WHILE** loops. In **FOR** loops, the elements that will be set to successive values of a vector or array are known before MATLAB begins execution of the **FOR** loop. In other words, in **FOR** loops, MATLAB knows how many times the code body of the **FOR** loop will be executed when it initiates execution of the **FOR** loop. To check your understanding of **FOR** loops, answer this question: How many times will a **FOR** loop iterate?

WHILE loops extend looping constructs to cases where MATLAB does *not* know in advance how many times the code body of the loop will be executed. Recursive user-defined functions in MATLAB allow a function to directly or indirectly call itself and, hence, provide a third way for the same code to be programmed to run multiple times.

WHILE loops are covered later in this chapter, in Section 8-5.

1. FOR loops are used in cases where you need more control over computations than allowed in cell-by-cell operations.

2. FOR loops start with a code line that begins with the MATLAB keyword `for` and end with a line that contains the MATLAB keyword `end`.

3. FOR loops iterate over a code block body using successive values of supplied vector or array.

4. If the intent is to have the loop variable in a FOR loop use successive elements of a vector, then a row vector must be used, and not a column vector.

5. If a FOR loop is supplied with an array, then successive values of the columns of the array are set to the value of the loop variable.

6. To understand a FOR loop, a good strategy is to "step through" the loop, which is of particular importance for nested FOR loops.

Synopsis for Section 8-3

8-4   Conditional Program Flow: IF–THEN–ELSE

In Section 8-3 and Section 8-5, you learned how to use the **FOR** and **WHILE** loop constructs that implement the general picture of iterative structures as discussed in Section 8-2 and shown in Figure 8-3. In this section, you will learn to use the MAT-LAB construct **IF-THEN-ELSE** conditional to implement the general picture of conditional structures as discussed in Section 8-2 and shown in Figure 8-2.

Look back at Figure 8-2. Conditional constructs of any kind depend primarily on your ability to form a valid relational/logical test in MATLAB. The whole point of a conditional is to direct MATLAB to execute different code paths through a program depending on the result of a test of current value settings. So, for you to use conditional code blocks in MATLAB effectively, you must effectively write the test on current values of variables. Review array relational and logic operators in Section 7-4 to be certain you have the required ability to form relational/logical tests.

We will describe the **IF-THEN-ELSE** construct and specific ways to use **IF-THEN-ELSE**. We will describe the use of the *positive-test-only* version of **IF-THEN-ELSE** and add to your knowledge of **IF-THEN-ELSE** by adding the *positive-negative* version of **IF-THEN-ELSE**. Finally, we will complete the topic by discussing the most general form of **IF-THEN-ELSE**, the *test-with-alternatives* version.

We will conclude by solving a second version of the inventory problem for Big League Bats.

8-4.1   General Forms of IF-THEN-ELSE

There is one general form for all **IF-THEN-ELSE** constructs, but we will consider the four specific subtypes of **IF-THEN-ELSE** shown in Figure 8-7. In each of the four forms, the bracketed statement *<commands>* means one or more standard MATLAB commands. Think of *<commands>* as representing a block of MATLAB code.

In each form, the first line begins with the MATLAB keyword `if`, and the last line is the MATLAB keyword `end`. As with **FOR** constructs, you should think of an **IF-THEN-ELSE** construct as one "super command" in MATLAB. That is, **FOR** and **IF-THEN-ELSE** have one point of entry. In the **IF-THEN-ELSE** case, the one entry is the `if` line, and the one exit is the `end` line.

The words **else** and **elseif** are also MATLAB keywords; we will discuss them shortly.

Examine the forms in Figure 8-7:

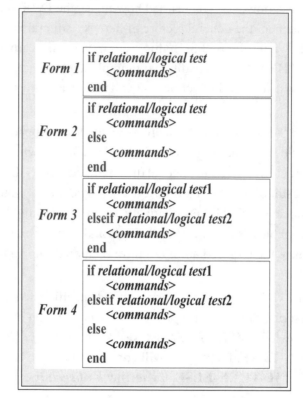

Figure 8-7: General Forms for IF-THEN-ELSE (Forms 3 and 4 can include any number of "elseif" clauses)

- **Form 1** is used when you want to perform a test, where if the test is **true**, then execute a block of commands. If the test is **false**, then none of the commands are executed.

- **Form 2** is used when you have two alternative paths, one to be followed if a test is **true** and one to be followed if that same test is **false**.

- **Form 3** is similar to Form 2 but with a subtle difference. Form 3 has two alternative paths, but there are two tests. If the first test is **true**, then commands immediately after it are executed. If the first test is **false**, then the second test is tried. If the result of the second test is **true**, then the code after it is run. If the second test is also **false**, then control passes out of the **IF-THEN-ELSE** block. The difference between Form 3 and Form 2 lies in how many tests are performed in the conditional.

Form 3 has a natural extension: Suppose you have three alternative paths, each one selected by a different test. Form 3 can accommodate that situation by adding another **elseif** block. Suppose you have 33 separate tests to perform, and a different set of commands for each. You could use Form 3 with 32 **elseif** clauses.

Note that the **if** and **elseif** blocks of Form 3 are executed in order, from the top of the **IF-THEN-ELSE** toward the bottom. The first test that results in **true** triggers execution of its associated statements. No other blocks will be executed.

- **Form 4** incorporates parts of Form 2 and Form 3. Form 4 has a test for the **if** clause and a test for the **elseif** clause, and if neither of the tests are true, the commands coming after the **else** clause will be run.

In the sections below, we'll give examples of each of the above forms of **IF-THEN-ELSE**.

8-4.2   IF-THEN-ELSE Form 1: *IF*

J&T Computers is a small company with 20 employees. The Human Relations Manager for J&T has convinced the CEO to give a $3,000 holiday bonus to everyone who works for the company provided all employees have a reasonable performance evaluation. Performance at J&T is evaluated on a scale from 0 to 5, with 5 indicating that the employee is to be promoted the following week, and 0 indicating that the employee is to be fired the following week. An acceptable performance rating is 3 or above.

Years ago, J&T switched to MATLAB to maintain employee performance and salary data. A portion of the J&T database is shown in Figure 8-8. The CEO has assigned the task to you, an entry-level programmer, to write a MATLAB program that determines the bonus employees are going to receive.

$$\begin{bmatrix} 99 & 3 & 45321 \\ 23 & 4 & 39278 \\ 105 & 5 & 73489 \\ 45 & 4 & 29340 \\ 122 & 2 & 41290 \end{bmatrix}$$

Figure 8-8: Performance Ratings and Salary Data for J&T: Column 1 has Employee Numbers, Column 2 has Performance Ratings, Column 3 has Salaries

The function below[9] accomplishes your task:

```
1  function raise = determineHolidayBonus(DB)
2  % determine holiday bonus for all employees:
3  %     if all employees perf rating >= 3
4  %        then bonus of $3000 is given
5  % INPUT: DB, array that contains in each row
6  %     <employee number> <perfRating> <salary>
7  % OUTPUT: raise, scalar either 0 or 3000
8
9     raise = 0;
10    if all(DB(:,2) >= 3)
11       raise = 3000;
12    end
```

Line 9 sets the value of the output variable, **raise**, to be zero. Lines 10-12 comprise a Form 1 version of **IF-THEN-ELSE**. If you do not remember the logical built-in function **all** (line 10), refer to MATLAB Help. The relational test in line 10 is a test over all elements in the second column of the database.

Why is the command in line 9 necessary? Make sure you can answer this question before going on.

8-4.3　IF-THEN-ELSE Form 2: *IF / ELSE*

Continuing with the problem from the previous section, suppose you run your function on the entire database of J&T employees, get the result, and take it to the CEO. The CEO has a change in heart. He has decided that every employee with a performance rating of 4 or 5 should get a 5 percent holiday bonus while all other employees should get a 2 percent bonus.

So, now your task is to write a new MATLAB function. This time, you should produce an output array having the first column as the employee number and the second column as the amount of salary each employee will receive for the holidays, including any holiday bonus the employee is to receive.

The CEO's problem specifications require that you need to examine the data for each employee separately. You could develop a solution for the new problem by using cell-by-cell operations, or you could develop a solution using a **FOR LOOP**. Assume that you opt to use a **FOR LOOP** and that you build the function below, which meets the

---

9. This example could be solved using `raise = 3000*allDB(:,2)>-3)`.

problem specifications.

```
1    function raiseA = determineHolidayBonus_2(DB)
2    % determine holiday paycheck for all employees:
3    % if employee performance rating == 4 | 5,
4    %        then bonus is 5% current salary
5    %        else bonus is 2% of current salary
6    % INPUT: DB, array that contains in each row
7    %          <employee number> <perfRating> <salary>
8    % OUTPUT: raiseA, array for all employees that contains in each row
9    %          <employee number> <adjustedSalaryAfterBonus>
10
11   raiseA = [];
12   for oneEmpData = DB'
13       empNum = oneEmpData(1);
14       empPerfRating = oneEmpData(2);
15       empCurrentSalary = oneEmpData(3);
16       if (empPerfRating == 4) | (empPerfRating == 5)
17           bonus = 0.05 * empCurrentSalary;
18       else
19           bonus = 0.02 * empCurrentSalary;
20       end
21       empNewSalary = empCurrentSalary + bonus;
22       raiseA = [raiseA; [empNum, empNewSalary]];
23   end
```

In line 11, you initialize the output variable to be an empty array. You need this step because in line 22 you build the output array row by row as you step through the loop. Without line 11, the first time you hit line 22 MATLAB would halt with an undefined variable error.

The **FOR LOOP** runs through elements of the input variable **DB** transposed. If you have trouble understanding why, review Section 8-3.3.

The **IF-ELSE** structure in lines 16-20 are an example of Form 2 in Figure 8-7. Line 16 includes a relational/logical test to check if the current employee record includes a performance rating of 4 or 5. If that test is **true**, the bonus for the current employee is set to 5% of the employee's salary. If that test is **false**, Line 19 sets the bonus to be 2% of the employee's salary.

The most important point to recognize in this section is that there is a match between the **IF-ELSE** form and the second version of the J&T problem specification. **IF-ELSE** is a form that requires a test. One set of actions is executed if the test returns a **true**, and a second set of actions is executed if the test returns a **false**.

8-4.4   IF-THEN-ELSE Form 3: *IF / ELSEIF*

Your job is not yet done. After you run your second function on the entire employee database and take the results back to the CEO, you find that again he has had a change of mind. This time, he tells you he was being overly generous before. He now wants you to set up the computation so that employees with performance ratings of 5 get a 4% bonus, those with performance ratings of 4 get a 2% bonus, and those with performance ratings of 3 get a 1% bonus. And this time he wants you to produce a listing of salary data only for the employees who will get a bonus.

You produce the following function to meet this revised problem specification:

```
1    function raiseA = determineHolidayBonus_3(DB)
2    % determine holiday paycheck for all employees:
3    %     if employee performance rating == 5,
4    %         then bonus is 4% current salary
5    %       elseif employee performance rating == 4,
6    %         then bonus is 2% current salary
7    %       elseif employee performance rating == 1,
8    %         then bonus is 1% current salary
9    % INPUT: DB, array that contains in each row
10   %     <employee number> <perfRating> <salary>
11   % OUTPUT: raiseA, array for ONLY the employees to get bonus
12   %                 first col is employee number
13   %                 second col is amount of salary with bonus
14
15    raiseA = [];
16    for oneEmpData = DB'
17        empNum = oneEmpData(1);
18        empPerfRating = oneEmpData(2);
19        empCurrentSalary = oneEmpData(3);
20        if empPerfRating == 5
21            bonus = 0.04 * empCurrentSalary;
22            empNewSalary = empCurrentSalary + bonus;
23            raiseA = [raiseA; [empNum, empNewSalary]];
24        elseif empPerfRating == 4
25            bonus = 0.02 * empCurrentSalary;
26            empNewSalary = empCurrentSalary + bonus;
27            raiseA = [raiseA; [empNum, empNewSalary]];
28        elseif empPerfRating == 3
29            bonus = 0.01 * empCurrentSalary;
30            empNewSalary = empCurrentSalary + bonus;
31            raiseA = [raiseA; [empNum, empNewSalary]];
32        end
33    end
```

This third version, **determineHolidayBonus_3**, is similar to the code for **determineHolidayBonus_2**. Other than changes in the comments, the differences lie in the conditional, lines 20-32, in **determineHolidayBonus_3**. The code here fits Form 3 of **IF-THEN-ELSE**, and we need to explain why. Form 3 in Figure 8-7 shows one **elseif** clause, but the code above has two **elseif** clauses. Though Form 3 includes only one **elseif** clause, you may use any number of **elseif** clauses you need in MATLAB. Think about the purpose for the **elseif** clause: It enables you to make a second test on current variables and to take appropriate action if the test returns **true**. You may have a number of alternatives in mind, so you may need to write many **elseif** clauses into a single Form 3 **IF-THEN-ELSE**.

There is a critical point to remember about Form 3 **IF-THEN-ELSE** blocks in MATLAB. The first **if** or **elseif** clause whose test is **true** is the <u>only</u> clause whose command block is executed. Look at the code below:

```
x = 82;
if x > 90
    y = 5;
 elseif x > 80
    y = 4;
 elseif x > 70
    y = 3;
 elseif x > 60
    y = 2;
end
```

The result of this code is that the variable **y** has the value 4. Only the command(s) after the first **true** test are executed. Be certain you understand this point since many beginning MATLAB programmers overlook it.

8-4.5 IF-THEN-ELSE Form 4: *IF / ELSEIF / ELSE*

After you run your third function, **determineHolidayBonus_3**, on the employee database and take the results back to the CEO, you find that once again he has changed his mind. Now he wants you to set up the computation so that employees with performance ratings of 5 get a 4% bonus, those with performance ratings of 4 get a 3% bonus, those with performance ratings of 3 get a 2% bonus, and everyone else gets a 1% bonus. Because all employees will get something, the output should include all J&T employees. It should also list the total amount the bonuses are costing the company.

You produce the following function to meet this revised problem specification:

```
1   function [totalBonuses, raiseA] = determineHolidayBonus_4(DB)
2   % determine holiday salary for all employees:
3   %     if employee performance rating == 5,
4   %         then bonus is 4% current salary
5   %       elseif employee performance rating == 4
6   %         then bonus is 3% current salary
7   %       elseif employee performance rating == 3
8   %         then bonus is 2% current salary
9   %       else
10  %         then bonus is 1% current salary
11  % INPUT: DB, array that contains in each row
12  %     <employee number> <perfRating> <salary>
13  % OUTPUT: totalBonuses, a scalar, total amount of bonuses
14  %          raiseA, an array containing all employees
15  %              first col is employee number
16  %              second col is amount of salary with bonus
17
18      raiseA = [];
19      totalBonuses = 0;
20      for oneEmpData = DB'
21          empNum = oneEmpData(1);
22          empPerfRating = oneEmpData(2);
23          empCurrentSalary = oneEmpData(3);
24          if empPerfRating == 5,
25              bonus = 0.04 * empCurrentSalary;
26          elseif empPerfRating == 4
27              bonus = 0.03 * empCurrentSalary;
28          elseif empPerfRating == 3
29              bonus = 0.02 * empCurrentSalary;
30          else
31              bonus = 0.01 * empCurrentSalary;
32          end
33          totalBonuses = totalBonuses + bonus;
34          empNewSalary = empCurrentSalary + bonus;
35          raiseA = [raiseA; [empNum, empNewSalary]];
36      end
```

Form 4 **IF-THEN-ELSE** is a combination of Form 2 (**IF-ELSE**) and Form 3 (**IF-ELSEIF**). **IF-ELSEIF-ELSE** forms have two or more test clauses followed by an **else** block of commands that will be run if none of the tests render **true**.

Create an array of the data shown in Figure 8-8:

```
>>empData = [99,3,45321; 23,4,39278; 105,5,73489; ...
             45,4,29340; 122,2,41290];
```

Calling the function above results in the following:

```
>> [totBonuses, adjustedSalary] = determineHolidayBonus_4(empData)
totBonuses =
        6317.4
adjustedSalary =
            99          46227
            23          40456
           105          76429
            45          30220
           122          41703
```

Form 4 performs each test (the **IF** test, then each of the **ELSEIF** tests) until one of them returns `true`. At that point, it executes the commands after the test that returned `true` and stops. Only one of the command blocks can be executed. If none of the tests return `true`, then the **ELSE** clause provides a backup set of commands to execute.

Implement each one of the four variations for computing the salary bonuses for employees at J&T and test each one. Be certain you understand the logic for each of the four forms of **IF-THEN-ELSE**, and be sure you understand which of the variations to choose for each problem setting.

Compare the code in Section 8-4.5 with the code in Section 8-4.4. The clearest difference is that the two functions have different purposes. Look especially at the code contained inside the **IF-THEN-ELSE** in `determineHolidayBonus_4`, and the **IF-THEN-ELSE** in `determineHolidayBonus_3`.

8-4.6   Solving the Big League Bats Problem: Version 2

Now return to the inventory problem for Big League Bats introduced in Section 8-1 and first solved using MATLAB in Section 8-3.5. Review the introduction and the solution using a **FOR LOOP**.

Here is the working part of the function we developed previously (`projectBigLeagueBats`):

```
for oneWeekSet = blbData'
  batSalesThisWeek = oneWeekSet(1);
  batsManuThisWeek = oneWeekSet(2);
  endInvThisWeek = startInvThisWeek+batsManuThisWeek-batSalesThisWeek;
  projEndInventory = [projEndInventory, endInvThisWeek];

  % update startInv to be the starting inventory for NEXT week
  startInvThisWeek = endInvThisWeek;
end
```

Now add a constraint to the problem that mirrors the Big League Bats corporate operation, which is to account for two more facts:

1. You do not want the manufacturing level to fall below 45,000 bats a week because that would leave too many of your employees on the shop floor idle.

2. You do not want the manufacturing level to go above 55,000 bats a week because that figure represents the upper limit of how many bats you can turn out per week without investing in more raw materials than the company can afford at present.

The following modification to the code will accomplish the task. The modification involves one of the four forms of **IF-THEN-ELSE**. Before you look at the code, see if you can determine which of the four forms fits this problem situation. Your ability to choose and apply the appropriate form of **IF-THEN-ELSE** is the key to your using **IF-THEN-ELSE** effectively for problem solving.

The code modification follows:

```
for oneWeekSet = blbData'
    batSalesThisWeek = oneWeekSet(1);
    batsManuThisWeek = oneWeekSet(2);
    if batsManuThisWeek < 45000
       batsManuThisWeek = 45000;
    elseif batsManuThisWeek > 55000
       batsManuThisWeek = 55000;
    end
   endInvThisWeek = startInvThisWeek+batsManuThisWeek-batSalesThisWeek;
   projEndInventory = [projEndInventory, endInvThisWeek];

% update startInv: the starting inventory for NEXT week
         startInvThisWeek = endInvThisWeek;
   end
```

In this case, the right form to use is **IF-ELSEIF**. Remember that you are required to produce no more than 55,000 bats, nor less than 45,000, regardless of your preferences. Those are fixed parameters. The **IF-ELSEIF** form allows multiple tests to be made.

8-4.7   Synopsis of MATLAB IF-THEN-ELSE Conditional

Along with the MATLAB **FOR LOOP**, the MATLAB **IF-THEN-ELSE** provides you with the tools you need to design and implement rather sophisticated programs. The **FOR LOOP** has several variations, but the fundamental form of the **FOR LOOP** remains largely the same regardless of the variation utilized. In contrast, the **IF-THEN-ELSE** conditional in MATLAB is best understood as four distinct forms: **IF,**

IF-ELSE, **IF-ELSEIF**, and **IF-ELSEIF-ELSE**. By understanding which problem settings match each of the four forms, using **IF-THEN-ELSE** becomes relatively straightforward.

1. **IF-THEN-ELSE can be used to express conditional program control. IF-THEN-ELSE is best understood in four distinct forms.**

2. **The first form is IF. In this form, one relational/logical test exists. During execution, if the test results in `true`, then the commands in the following block are run. If the test results in `false`, then the commands are not run.**

3. **The second form is IF-ELSE. This form performs a relational/logical test and, if `true`, then runs a set of commands. If `false`, an alternative set of commands is run.**

4. **The third form is IF-ELSEIF. There are two essential points to remember: (a) there can be multiple ELSEIF clauses and (b) only one (at most) code block following a test will be run, which will be the one following the first test that results in `true`.**

5. **The fourth form is IF-ELSEIF-ELSE. This form is a combination of the second (IF-ELSE) and third (IF-ELSEIF) forms.**

6. **The key to effective use of IF-THEN-ELSE conditionals is to correctly match the problem situation you have with one of the appropriate four forms.**

Synopsis for Section 8-4

8-5   Iterative Program Flow: While LOOP

In Section 8-3 you learned to apply the MATLAB **FOR** loop to problems that involved solving a set of problem instances where solving one instance depended on the solution to previous instance – like the Big League Bats inventory problem. The type of iteration you learned in Section 8-3 (**FOR**) depended on knowing how many times you were going to "iterate over" a code block. For example, in the Big League Bats problem, you knew the number of weeks over which you wanted to run the inventory simulation.

In Section 8-4 you learned to write programs with conditional statements using **IF** constructs. You saw that relational and logical tests provided the key mechanism to allow you to develop code that can take different paths depending on the values stored in program variables.

In this section you will see a combination of these two capabilities: iterative programming and conditional programming. Often, there is a need in program development for this combination. A common use of a **WHILE** loop is to rerun a sequence of code again and again (iterative programming) as long as some test on internal variables is **TRUE** (conditional programming). Unlike standard use of a **FOR** loop, with a **WHILE** loop you cannot generally predict how many times the code sequence will be re-run before the loop starts execution. It is this distinction – that with a **FOR** you know how many times the loop will iterate before the loop is entered, but that in a **WHILE** you do not know how many times the loop will iterate – that you need to put firmly in your mind because that single characteristic suggests which tool (**WHILE** or **FOR**) you should apply in a given problem.[10]

In Section 8-5.1 we will describe the general form of a **WHILE** loop. In Section 8-5.2 we will go over an example to illustrate how a **WHILE** loop is used to implement iteration. In the example Section 8-5.2 we will also point out the key problem you have to keep in mind if you use **WHILE** loops: unlike a **FOR** loop, a **WHILE** loop is *not* guaranteed to every complete.

---

10. In reality, any **FOR** loop can be written as a **WHILE** loop, and any **WHILE** loop can be written as a **FOR** loop. Converting a **FOR** to a **WHILE** is straightforward. Converting a **WHILE** to a **FOR** is possible provided the computer language you are using gives a way to "break out" of the execution of a loop. MATLAB supplies two built-in functions for that purpose: `return` and `break`. A good exercise is to take a simple **FOR** loop and convert it a **WHILE** loop, and similarly to convert a simple **FOR** loop to a **WHILE** loop. The reason for having both **FOR** and **WHILE** available is that some problems are easy to understand as **FOR** situations, while others are easy to understand as **WHILE** situations. Matching the best tool (**WHILE** or **FOR**) to a given problem that requires iteration is one of your central tasks for this chapter.

8-5.1   General Form of a WHILE Loop in MATLAB

The general form of a **WHILE** *loop* in MATLAB is shown in Figure 8-9. The parts of Figure 8-9 shown in italics are what you must fill in with specifics in the loop you are building.

```
while relational_or_logical_test
    Command_1
    Command_2
    • • •
    Command_N
end
```

Figure 8-9: General Form of a WHILE Construct in MATLAB

You could describe a **WHILE** loop this way:

> *A* **WHILE LOOP** *is a block of MATLAB code starting with the MATLAB keyword* **while** *and ending with the MATLAB keyword* **end** *and enclosing one or more MATLAB commands. On the same code line as the keyword* **while**, *there must be a condition that, when* **FALSE** *will terminate iteration.*

The crux idea of a **WHILE** loop is that you have a test that determines when a block of MATLAB commands should be executed. As long as the test gives a value or **TRUE**, you continue to run the block of commands. As soon as the test gives a value of **FALSE**, you stop running the block and go on to the next MATLAB command after the **end** marker of the **WHILE** loop.

8-5.2   Example of WHILE Loop

Suppose you have just completed a summer internship at a Silicon Valley engineering company, Big Chip Enterprises. And you have managed to save $10,000 over the course of the summer. (You had to work *really* long hours so you didn't "get out" much. But the pay was great!) In considering how to use your money, you decide to take $2,000 for mad money, and use the remaining $8,000 to start a savings account.

You've had a long time dream of taking a world cruise, so you decide to use that savings account to target a goal of building up $25,000 for a really high class world cruise when you graduate. Your question: How many years will it take you to save the other $17,000? First, you have to structure the problem: what do you need to know?

what are the inputs to your problem? what are the outputs you want from your solution?

Your base question is "How many years …?" Another output that would be useful is the amount saved beyond the target you set ($25,000) – its unlikely after a whole-number of years that you would hit your goal exactly. There are a few other parameters of the problem you have to pin down to create a well structured problem spec. Suppose you assume that the interest rate is stable at 5%, that you make a regular deposit at the end of every year of $2,000, and you already established that your initial deposit would be $8,.0000

Why not just use a **FOR** loop to implement a solution? The reason is that you don't know how many years you have to save, and to use a basic **FOR** loop solution you need to know how many years you want to run the savings program. Maybe there is a related problem that could be solved with a **FOR** approach?

It turns out … there is a related problem you could easily solve with a **FOR** loop: How much would your savings accumulate after 10 years? (Or any other number of years you choose…) This problem has an straightforward solution assuming a given interest rate, a given initial deposit, and a given amount you deposit each year. The MATLAB function below will do the trick.

```
function balance = amountSavedAfterNyears ...
                     (initialDeposit, intRate, yearlyDeposit, yrNum)
% computes amount saved after a fixed number of years
% INPUT: initialDeposit, scalar, amount deposited account
%                            at time = 0 yrs
%        intRate, scalar, decimal rate of simple annual interest
%        yearlyDeposit, scalar, amount added to account at
%                            time = 1 year, time = 2 years, etc
%        yrNum, scalar, number of years for money to accumulate
% OUTPUT: balance, scalar, amount saved after yrNum years
% sample call
%        bal = amountSavedAfterNyears(8000, 0.05, 2000, 10)

    balance = initialDeposit;
    for yrCounter = 1:yrNum
        intEarned = balance * intRate;
        balance = balance + intEarned + yearlyDeposit;
    end
```

How can you use **amountSavedAfterNyears** to answer your initial question: How many years will it take to save …? You *could* solve the number-of-years problem using the solution to the how-much-after-N-years problem … but it would be tedious. At the outset, your balance is less than your goal. So you need to wait until 1 year passes, then see where you stand. If you call **amountSavedAfterNyears** with the *yrNum* parameter set to 1, you find…

```
>> bal = amountSavedAfterNyears(8000, 0.05, 2000, 1)
bal =
      10400.00
```

Now you can test again to see if you have reached goal. After 1 year you are still under $25,000, and thus you need to continue saving for another year (at least)…

Now you continue the savings plan for another year, and then at the end of 2 years you find …

```
>> bal = amountSavedAfterNyears(8000, 0.05, 2000, 2)
bal =
      12920.00
```

Still under $25,000. Again, you have to save for one more year (at least) …

```
>> bal = amountSavedAfterNyears(8000, 0.05, 2000, 3)
bal =
      15566.00
```

Still under $25,000, so you must continue the same steps.

You can see where the progression is going, and how the *process* is leading you step by step to …

1. test to see if the your current balance is less than your goal

2. a) if it is less, then continue the savings plan for another year

   b) if the current balance is not less than the goal, then you have found the number of years you must save to meet your savings goal and you can terminate the iteration.

Now … look at the general description of a **WHILE** loop shown in Figure 8-9. Even though the process we went through above using **amountSavedAfterNYears** does not quite match the general description of a **WHILE** loop, there is a remarkable similarity. The most important facet of that similarity lies in the cycle of *test, then run code again if the test is true*. The heavy lifting in our process above was manual; you ran **amountSavedAfterNYears** for successive year, and tested for each year if your saving goal was reached. But by using a **WHILE** loop solution, we can automate the test/run cycle.

The MATLAB function below implements a **WHILE** solution that solves the original question: How many years …?

```
function [yrCounter, balance] = computeYrsToGoal(initialDeposit, ...
                                intRate, yearlyDeposit, goal)
% computes minimal number of years to reach a savings goal
% INPUT: initialDeposit, scalar, amount put in account at time = 0
%        intRate, scalar, decimal rate of simple annual interest
%        yearlyDeposit, scalar, amount added to account at
%                                year = 1, year = 2, etc
%        goal, scalar, amount that is the savings goal
% OUTPUT: yrCounter, scalar, number of years necessary to achieve
%                            goal
%        balance, scalar, number of years that is the solution
% sample call
%        [yrGoal, bal] = computeYrsToGoal(8000, 0.05, 2000, 25000)

    balance = initialDeposit;
    yrCounter = 0;
    while (balance < goal)
        intEarned = balance * intRate;
        balance = balance + intEarned + yearlyDeposit;
        yrCounter = yrCounter + 1;
    end
```

Examine critically the code inside the **WHILE** loop in **computeYrsToGoal**. Step through the each iteration by hand systematically and fill in the following table …

At start of loop: yrNum	At start of loop: balance	At end of loop: intEarned	At end of loop: balance	At end of loop: yrNum
0	?	?	?	?
1	?	?	?	?
2	?	?	?	?
3	?	?	?	?
4	?	?	?	?
5	?	?	?	?
6	?	?	?	?
7	?	?	?	?

When you call **computeYrsToGoal** with the problem you have set up (to save $25,000), you find …

```
[yrGoal, bal] = computeYrsToGoal(8000, 0.05, 2000, 25000)
yrGoal =
        7.00
bal =
    27540.82
```

To check this before we go further, call **amountSavedAfterNyears** with the number of years set to 7, and see if the result agrees with the result from **computeYrsToGoal**. Also, check with the number of years set to 6. (What is the purpose of the test for the number of years set to 6 of value? We already know from the test for number of years set to 7 that the goal is reached for that input.)

In some ways, a **WHILE** loop is easier conceptually than a **FOR** loop. Go back and compare the abstract descriptions of the two different loop types as shown in Figure 8-4 (**FOR**) and Figure 8-9 (**WHILE**). The **FOR** loop gives you machinery to step through a vector (or array). But in a **WHILE** loop you must set up how the variables inside loop change. In **computeYrsToGoal** notice that the *yrNum* variable must be incremented explicitly, but in **amountSavedAfterNyears** the *yrNum* variable is a loop variable of the **FOR** loop, and that changing (incrementing) the value of *yrNum* on successive passes through the iteration is handled "automatically" by the mechanism of the **FOR** loop itself.

There is one additional, and very important difference between **FOR** loops and **WHILE** loops. A FOR loop will terminate after it has stepped through all of its iteration values as given in the first line of the **FOR** statement. For example, if the first line of a **FOR** loop is …

```
for i = 1:32
```

then *i* will take on 1, then 2, then   , and finally the iteration will stop after the pass that starts with *i* taking on the value of 32. This is hard core predictable!

But now consider a **WHILE** loop that starts with

```
while y < 10
```

The **WHILE** loop that starts with this line will continue until the value of *y* is less than 10. But suppose the **WHILE** loop, and its set up, are as follows …

```
y = 11;
while y > 10
    y = y + 1;
    z = 22*y^2.5;
end
```

How long will it take for this **WHILE** loop to complete its job and terminate? Answer… it will *never terminate*. This small example shows the problem. Although in some ways **WHILE** loops are more flexible that **FOR** loops, the price you pay is that you must be careful when writing a **WHILE** loop that it does not become a "run away" loop.

One way to guard against run away **WHILE** loops is to insert temporary pauses in the loop during each iteration, and allows you to figure out if the loop is doing what you intended. For example, using the **WHILE** loop code from above …

```
function testRunaway
    y = 11;
    while y > 10
        y = y + 1;
        z = 22*y^2.5;


        y, z
        pause
    end
```

The two lines of code before the **end** statement have the effect of (a) displaying the values of x and y on the command window and (b) suspending execution until any key on the keyboard is struck. Inserting those two extra lines of code will allow you to follow the iteration steps your code is running one step at a time. Once you see that the code is doing what you want – or that its not doing what you want – then execute a <cntl>C to break out of the execution.

An even better way to test your **WHILE** loops is to apply the methods you learned in Section 4-4 for debugging MATLAB functions. Figure 8-10 shows an edit window up on a function containing the code above. Note the "red flag" stop that has been placed in the code to halt execution at the end of each iteration. When stopped, what you would normally do is to examine the values of x and y and, like above with the use of **pause**, you would track to be sure the code is not going to run away. Once convinced, you could remove the red flag.

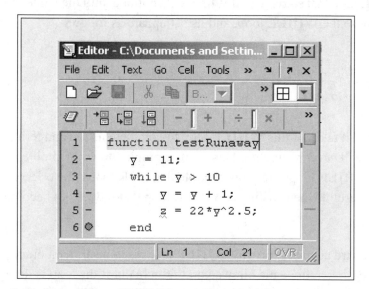

Figure 8-10: Debugging a WHILE loop: Watching for "Run Away" Conditions

8-5.3 Synopsis of MATLAB WHILE Loop

Along with the MATLAB **FOR LOOP**, the MATLAB **IF-THEN-ELSE** provides you with the tools you need to design and implement rather sophisticated programs. The **FOR LOOP** has several variations, but the fundamental form of the **FOR LOOP** remains largely the same regardless of the variation utilized. In contrast, the **IF-THEN-ELSE** conditional in MATLAB is best understood as four distinct forms: **IF**, **IF-ELSE**, **IF-ELSEIF**, and **IF-ELSEIF-ELSE**. By understanding which problem settings match each of the four forms, using **IF-THEN-ELSE** becomes relatively straightforward.

---

1. **WHILE loops are used in cases where you need more control over computations than allowed in cell-by-cell operations. and at the point where MATLAB starts the WHILE loop, the number of times the loop will iterate is not known.**

2. **WHILE loops start with a code line that begins with the MATLAB keyword `while` and end with a line that contains the MATLAB keyword `end`.**

3. **WHILE loops iterate over a code block body until a relational/logical test in the first line of the WHILE statement evaluates to FALSE.**

4. **Care must be exercised to prevent "run away" WHILE loops.**

Synopsis for Section 8-5

8-6 Problem Sets for IF-THEN-ELSE, For LOOP, and WHILE LOOP

Just a note before you begin these problems. Be careful with the problems involving loops in this problem set, particularly the **WHILE** loops. If you do not use the **WHILE** structures properly, MATLAB may get caught in an infinite loop. If MATLAB is taking too long running a function, click in the Command Window and press <cntl>C (hold down the control button and hit the C key) to interrupt MATLAB from running.

8-6.1 Set A: Nuts and Bolts Problems for Programming Constructs

Problem 8-A.1 (Section 8-4.5)

Write a MATLAB function, **Prob8_A_1**, that inputs a scalar **x** and returns a scalar **y** according to the following rules. Use Form 4 of **IF-THEN-ELSE** in your function.

$$\left.\begin{array}{l} y = -x - \pi \\ y = \sin(x) \\ y = -x + \pi \end{array}\right\} for \left\{\begin{array}{l} x \leq -\pi \\ -\pi < x \leq \pi \\ \pi < x \end{array}\right.$$

Problem 8-A.2 (Section 8-4.5)

Create a function, **Prob8_A_2**, that returns the course grade for a given numerical score for the term. Numerical scores are on a 100-point scale, and final course grades are on a 4-point scale. Your function must input a numerical score and return a vector containing the numerical score for a student in the first position and the corresponding course grade in the second position. Here is the grading scale for the course:

- 90 and above: 4.0
- Between 80 and 89: 3.0
- Between 70 and 79: 2.0
- Between 60 and 69: 1.0
- 59 and below: 0.0

For example, if the input score is **84**, your function must return the term grade **3**. Again, use Form 4.

Problem 8-A.3 (Section 8-4.5)

Write a function, **Prob8_A_3**, using two scalar inputs **x** and **y** and the following code:

```
if x > 3
   if y > x
      disp('x is greater than 3 and y is greater than x')
   end
end
```

This code is two nested Form 1 versions of **IF-THEN-ELSE**.

Modify your function and save it as **Prob8_A_3b**, so you have one Form 1 conditional.

Test your revised code with the following pairs of **x** and **y**:

x	1	1	4	4	4
y	2	4	3	4	5

Problem 8-A.4 (Section 8-4.2 and Section 8-4.3)

Are the following Forms 1 and 2 **IF-THEN-ELSE** statements functionally equivalent? That is, do they generate the same result? Why?

Code 1	Code 2

`if x > y`
`   z = 1;`
`else`
`   z = 0;`
`end`
	`z = 0;`
	`if x > y`
	`   z = 1;`
	`end`

Problem 8-A.5 (Section 8-4.5)

Write a MATLAB function Form 3 conditional as a function, **Prob8_A_5**, that inputs an angle in degrees and returns one three-element vector containing the quadrant of the angle (consider angles between 0 and 360 degrees) and returns the sine and cosine of the angle. See the following drawing for quadrants.

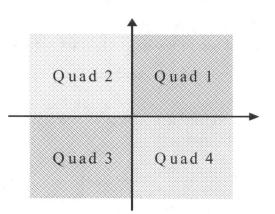

Assume that 0° is in quadrant 1, 90° is in quadrant 2, 180° is in quadrant 3, and 270° is in quadrant 4.

Use Form 4 of **IF-THEN-ELSE** for this problem.

Problem 8-A.6 (Section 8-3.2)

Do the following tasks using loop structures:

1. Find the sum of the first 50 integers.

2. Find the product of the first 10 integers.

Problem 8-A.7 (Section 8-3.2 and Section 8-4.5)

Consider the Form 4 **IF-THEN-ELSE** statement shown below and answer the following questions. To get your answers, step through the code the way MATLAB does when the code is executed. Keep track of the variables and how their values change each time through the loop. It is easiest to do this if you write down the variables and their values in a simple table. After you have completed the paper and pencil exercise, perform this activity using MATLAB, and verify your answers.

```
x=140;
for i = 1:10
   if x < 1
      break
   elseif x>50
      x=x/5;
   elseif x>60
      x=x/7;
   else
      x=x/10;
   end
end
disp(x)
```

The built-in function **break** is serving a special purpose in the code above. Look up **break** in MATLAB Help and understand how it is being used here.

1. What will MATLAB display?
2. How many times will this loop be executed?

Problem 8-A.8 (Section 8-3.2 and Section 8-4.2)

Consider the following code:

```
a = [];
x = 0;
for i = 1:4
    y = 3*x^2+3;
    if y < 10
        z = y;
    end
    x = x + 1;
    a = [a; [x,y,z] ];
end
```

This code implements a **FOR LOOP** that updates the values of x, y, and z. First, implement the code on a piece of paper and fill in the following table. Next, implement the code in a function file, **Prob8_A_8**, run it, and verify your answers. a is used to record values of x, y, and z at the end of each pass.

Pass	Value of x	Value of y	Value of z
First	?	?	?
Second	?	?	?
Third	?	?	?
Fourth	?	?	?

Problem 8-A.9 (Section 8-3.3)

Write a MATLAB function, **Prob8_A_9**, that inputs two vectors and computes and returns a vector that contains their cell-by-cell sum. Assume that inputs are two same-length vectors and use a **FOR LOOP** over columns to compute the cell-by-cell sum.

Problem 8-A.10 (Section 8-3.4)

Write a MATLAB function, **Prob8_A_10**, that inputs two arrays and computes and returns an array that contains their cell-by-cell sum. Assume that inputs are two same-sized arrays and use two nested **FOR LOOP** constructs to compute the cell-by-cell sum.

Problem 8-A.11 (Section 8-3.2)

Consider this sequence:

$$7k^3 \quad k = 1, 2, 3, \ldots, 10$$

Your task is to compute the sum of the first ten elements of the series.

Write a function, **Prob8_A_11a**, that solves the problem using vector operations.

Write a second function, **Prob8_A_11b**, that solves the problem using a **FOR LOOP**.

Be sure that your two functions are returning the same answer.

Problem 8-A.12 (Section 8-3.2)

For the sequence given in problem A.11, write a function, **Prob8_A_12**, that inputs a positive integer **N** and calculates and returns a vector containing the first **N** series sums as a vector output. (Use the series in Problem A.11.) Use **Prob8_A_11b** (the second function you wrote for Problem A.11) in your new function, **Prob8_A_12**.

Problem 8-A.13 (Section 8-3.2)

Create a function, **Prob8_A_13**, that inputs a vector and outputs the maximum value in the input vector. Use a **FOR LOOP** instead of the built-in function **max**.

Problem 8-A.14 (Section 8-3.2)

A student writes the following code as part of a function to calculate the growth of $10,000 in a bank account over the next 25 years. He assumes a constant interest rate of 6 percent:

```
initial_money=10000 % dollars
for year = 1:25
    % new balance is found by applying the interest
    money=(initial_money*1.06);
    % increment year by one
    year=year+1;
end
```

Put this code into a function that has no input and that returns **money,** and call it **Prob8_A_14**.

This code has two runtime errors. Find and fix the problems and test your revised function in MATLAB.

Problem 8-A.15 (Section 8-3.2)

George opens a new bank account with $35,000. He is planning to deposit $5,000 every year for the next 14 years. The interest rate on his account is 4 percent and is constant. George would like to create a plot showing the growth of his money. He would like enough flexibility to see what would happen with differing values for initial deposit, annual deposits, interest rates, and length of time he will maintain the account.

Create a function, **Prob8_A_15**. The function should input the four variables that George wants to vary, create a plot of the amount of money in George's account versus time, and output a vector where each element is the starting yearly balance for the corresponding year.

Problem 8-A.16 (Section 8-5.2)

Use a **WHILE** loop to determine how many terms in the series are required for the sum of the terms to exceed 5000:

$$2^k \quad k = 1, 2, 3, \ldots$$

Create a function, **Prob8_A_16,** that inputs the target sum for the series and returns the number of terms in the series.

Problem 8-A.17 (Section 8-5.2)

For the sequence given in problem A.16, write a function, **Prob8_A_17**, that inputs a positive integer **M** and calculates and returns a vector containing the first **N** series sums that are LESS THAN M.

Problem 8-A.18 (Section 8-3.6, Section 8-5.3)

Write four functions to do the following tasks:

1. Find the sum of the first 100 EVEN integers.
2. Find the sum of the first 100 ODD integers.
3. Find the number N, for which the product of integers from 1 to N (N factorial) exceeds 10,000.
4. Find the number N, for which the product of EVEN integers from 2 to N exceeds 10,000.

Decide on the most appropriate inputs and outputs for your functions. Choose the most appropriate **LOOP** structure to do each task.

Problem 8-A.19 (Section 8-5.2)

Consider the following code:

```
a = 50;
b = 30;
while a > b
    if a-b > 5
        z = -8;
    else
        z = 5;
    end
    a=a+z;
end
```

Implement the code on paper and fill in the following table. Then, implement the code in a function, **Prob8_A_19**, run it, and verify your answers.

Pass	Value of a	Value of b	Value of z
1st	?	?	?
2nd	?	?	?
3rd	?	?	?
4th	?	?	?
5th	?	?	?
6th	?	?	?

Problem 8-A.20 (Section 8-5.2)

Write a function, **Prob8_A_20**, that returns the value of **N** such that the series

sum for $\sum_{k=1}^{\infty} \frac{1}{k^2}$ is precise to within 0.001 and returns the value of the series to that

precision.

Since this infinite series is monotonically decreasing, you can determine when to stop by comparing the sum after **M** terms have been added to the sum after **M-1** terms have been added:

```
sum_M-sum_M 1 < 0.001, then N = M.
```

This problem is common when dealing with computing an infinite series. You cannot set up your computational loop to do an infinite number of iterations through the loop because that would require infinite time. The solution is to determine a degree of precision that you will be content with, stop the loop, and return the answer when that level of precision is reached.

8-6.2   Set B: Problem Solving Using Programming Constructs

Problem 8-B.1

The *Freshest Foods Market* is announcing a new promotion. This is their theme:

> "Cutting down prices as you buy! 10% off every additional item you buy after your purchase totals $20, and 20% off after $50!"

When she reaches the counter, Julie arranges her groceries as follows:

Item	Gum	Orange	Cereal	Cheese	Coffee	Tomato	Pizza	Sushi	Shrimp
Unit Price ($)	0.45	2.95	1.99	4.99	4.95	3.99	7.99	16.95	11.99
Number of items	1	1	2	1	1	2	2	1	2

How much will Julie pay? Build a function, **Prob8_B_1**, to answer the question. Input should be an array representing data as in the table above. Output should be the total bill Julie will pay, as a scalar.

Use a **FOR** loop in your function. Consider how you would solve this problem with vector operations using the built-in function **cumsum** as part of your solution.

Problem 8-B.2

David deposits $1,000 at Bank A and Mark deposits $1,000 at Bank B. Bank A offers a 5 percent annual interest rate whereas Bank B offers 8.5 percent.

Write a MATLAB function, **Prob8_B_2**, that inputs David's initial deposit, Mark's initial deposit, the interest rate (as a percentage) offered by David's bank, and the interest rate (as a percentage) offered by Mark's bank.

Your function should return the number of years for Mark's savings grow to be at least twice as much as David's.

Solve this problem using the **FOR** and **WHILE** loops. Look up **return** to implement in the **FOR LOOP**.

Assume that the annual interest payments are deposited to the accounts at the end of each year and that David and Mark do not deposit any additional money after their initial deposit.

Problem 8-B.3

This following code line will generate a vector of 100 random numbers in the range [–0.5, 0.5]. (Look up **rand** in MATLAB Help.)

```
My_array=rand([1 100])- 0.5;
```

Build a function, **Prob8_B_3**, that inputs a vector of 100 random numbers between -0.5 to 0.5. Your function should scan the input vector from beginning to end and count the number of times the sign changes moving from one element to the next. Consider zero to be neither positive nor negative. Hence, you must consider three number ranges: less than zero, equal to zero, and greater than zero.

You might want to remind yourself of the answers to the following two questions: When is the product of two numbers negative? When is the product zero?

Problem 8-B.4

Write a game function, **Prob8_B_4**, that randomly chooses an integer between 0 and 100 and asks the user to guess the number. The program should prompt "Guess the number!" until the user enters the correct number or a number outside of the range. If the guess is correct, the program displays "Good guess!" and quits. If not, the program prompts "Guess higher" or "Guess lower" according to the number guessed.

When the program terminates, the program displays the number of attempts made by the user.

Use the built-in function **input** to get input from the player. (See MATLAB Help.)

Problem 8-B.5

The Fibonacci numbers are defined by the sequence such that each value is the sum of the previous two values. The first two numbers in the Fibonacci sequence are given below:

$$F_1 = 1$$
$$F_2 = 1$$

The formula for calculating the successive numbers in the sequence is the following:

$$F_i = F_{i-1} + F_{i-2}$$

where i is an integer greater than 2.

Write a function, **Prob8_B_5**, that takes one argument, a positive integer N, and outputs a vector of Fibonacci values so the last number in the sequence is the greatest Fibonacci number that is less than N.

Problem 8-B.6

Your computer science professor gives you an extra credit project. Your job is to write a function, **Prob8_B_6**, that determines the end-of-term grade for each student in the course. The input to your function will be an array of student grades. The professor gives you the following sample input:

Student ID	Homework	Midterm	Term Project	Final
39	61	73	85	64
271	95	79	80	100
418	72	65	94	81
665	83	86	82	92
860	25	40	93	85

All grades will be weighted equally in determining the final grade:

$$overallGrade = \frac{homework + midterm + termpProj + final}{4}$$

The professor promises the students to take individual progress into account at the end of the term. If the term project grade is at least 15 points higher than the homework grade, and the final grade is at least 10 points higher than the midterm grade, you must calculate the overall grade by ignoring their homework and midterm scores:

$$overallGrade = \frac{termpProj + final}{2}$$

Your function takes an $N \times 5$ array and outputs an $N \times 2$ array. The first and second columns in the output should be the student ID and end-of-term grade, respectively. N is a variable, i.e., your function should work for any number of students.

Problem 8-B.7

A mathematical sequence, f, is defined this way:

$$f(n) = f(n-1) + f(n-1) \times f(n-2) \quad \text{for } n > 2,$$

$$f(1) = 1$$

$$f(2) = 1$$

Write a function, **Prob8_B_7**, that inputs a positive integer N as input and outputs the first N numbers in this sequence as a vector.

Test your function with the following:

- For an input of 1, the output should be **1**
- For an input of 2, the output should be **[1 1]**
- For an input of 3, the output should be **[1 1 2]**
- For an input of 4, the output should be **[1 1 2 4]**

Problem 8-B.8

The half-life of a radioisotope is the time required for half of the radioactive mass in any sample to undergo radioactive decay. In a radioactive sample, after one half-life, half of the original sample remains. After three half-lives, one-eighth of the original mass remains.

Write a MATLAB function, **Prob8_B_8**, that determines how many years it would take for N atoms of radioactive material, with half-life of HL years, to be reduced to a level at or below the target threshold of T atoms. (Normally, we would use the mass of the radioactive material, but here we will use the number of atoms.)

This function will have three inputs:

1. The total number of radioactive atoms we start with (integer)
2. The length of the half-life (in years)
3. The target threshold number of atoms (integer)

It will have two outputs:

1. **numYears**: the total number of years it will take for the radioactive material to decay to a level at or below the target threshold
2. **atomsLeft**: the number of atoms left after **numYears**

Problem 8-B.9

As the ecological supervisor for the Great Hitonkin Forest, you are contracted by a lumber company to find out the number of trees it will be allowed to cut down over the coming years. The number of trees a lumber company is allowed to chop in a given year can be modeled as 0.5 times the number of seedlings planted the previous year minus 0.1 times the number of trees chopped down the previous year.

The company provides you with the following data on the number of seedlings it will plant over the next 10 years:

```
seedlingPlanted =
[2700, 3000, 3000, 3200, 4000, 3700, 3300, 4200, 4400, 4100]
```

The lumber company tells you it plans to chop down 3,000 trees this year.

Write a MATLAB function, **Prob8_B_9**, that inputs the following:

- the number of trees chopped down in the first year
- a vector for the number of seedlings that will be planted in the next 10 years

Your function should plot the projected number of trees the lumber company is allowed to chop down in the next 10 years. (Label your plot well because it will be part of your report to the lumber company.)

Problem 8-B.10

Elaine opens a savings account with a 6 percent interest rate and deposits $10,000. In the next 30 years, Elaine does the following:

1. She deposits $10,000 at the beginning of every year.
2. In years when she has more than $100,000, she deposits $10,000 and then withdraws 10 percent of her total balance to buy mutual funds.

Write a MATLAB function, **Prob8_B_10**, that has no inputs or outputs. The function should plot Elaine's account balance over the next 30 years. Do not include the value of her mutual funds.

Problem 8-B.11

The Registrar's Office wants you to project the number of graduating students in the coming years. The total number of graduating students can be modeled this way:

> The number of senior students who did not fulfill their requirements in the previous year (we are going to assume that they finish in one additional year)

PLUS

the number of new enrollments in senior class in the current year

MINUS

the number of senior students who did not fulfill their requirements in the current year.

The Registrar's Office gives you the projected new enrollment in the senior class and the projected number of students that will not fulfill their requirements. These projected data are for each of the next 15 years:

```
projectedEnrollment =
[95, 81, 76, 88, 96, 121, 133, 112, 91, 89, 79, 74, 91, 121, 75]

projectedStudentsNotFulfilling =
[5, 6, 3, 10, 7, 4, 6, 8, 8, 3, 4, 3, 2, 10, 8]
```

The Registrar indicates that a total of 90 students graduated at the end of the first year. Write a MATLAB function, **Prob8_B_11**, that inputs the **studentsGraduatingYear1**, the **projectedEnrollment**, and the **projectedStudentsNotFulfilling**. It should have no output variables. Your function should plot the projected number of students graduating over the next 15 years.

Problem 8 B.12

Your engineering expertise is requested by the Vandelay Truck Company. Truck prices are projected to increase next year, so Vandelay would like to keep at least 1,000 trucks in the warehouse until the prices go up. The capacity of the warehouse is 1,500 trucks and is never to be exceeded. The following model for warehouse management has been developed:

- The production schedule for the next twelve months is **[130 90 105 152 174 146 168 194 203 185 164 191]**. All manufactured trucks will be stored in the warehouse until they are sold.
- Vandelay is planning to sell 15 percent of inventory once more than 1,050 trucks are in the warehouse. Otherwise, no trucks will be sold.
- Current inventory is 850 trucks.

Vandelay would like to know if inventory will fall below 1,000 or exceed 1,500. Create a plot of inventory over the next 13 months.

For quick and easy plot generation, write a MATLAB function, **Prob8_B_12**. Your function should input the percentage of trucks that will be sold (15 percent in this case) when inventory goes over 1,050 and should output a vector of inventories. It should plot a graph of inventory for the next 13 months.

Assume that the trucks produced in one month are transferred to the warehouse at the beginning of the next month. Start with a beginning inventory of 850 and run your simulation from month 2 to month 13.

Hint: 15 percent of inventory will not always be an integer. Use the built-in function **round** to round the numbers to the nearest integer.

Problem 8-B.13

Dr. Crane opens a bank account with an initial deposit of $2,125. He will receive 6.25 percent interest, compounded quarterly. Right after he deposits the money, he moves to Paris and stays there for 19 years. He does not make any deposits or withdrawals during this time.

Nineteen years later, he returns home and goes to collect the money from his account. How many dollars in interest has his account earned over the last 19 years? (Interest is compounded four times every year, which you can assume means that for every three months, 6.25%/4 interest is paid to Dr. Crane's account.)

Write a MATLAB function, **Prob8_B_13**, which will take the initial deposit, number of years, and the interest rate (in percentage) as inputs. The only output will be the interest he earned over the years he was gone.

Problem 8-B.14

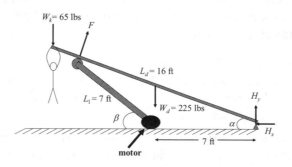

The sketch above shows an electric motor-powered underground garage door. The door weighs 225 pounds and is 16 feet long. The lever that lifts the door is 7 feet long and has negligible weight. A child hangs on the tip of the door as the door moves up.

The magnitude of torque produced by the electric motor can be calculated with the following:

$$\tau = F \times L_l \times \mathrm{Sin}\left(90 - \frac{\beta}{2}\right)$$

F is the force applied on the door by the lever and L_l is the length of the lever.

We can calculate the force F by calculating the moments around the hinge:

$$F = \frac{\left(W_d \frac{L_d}{2} + W_k L_d\right)\cos^2\left(\frac{\beta}{2}\right)}{L_l(1 + \cos(\beta))}$$

W_d and L_d are the weight and length of the door, respectively, W_k is the weight of the child hanging on the door, L_l is the length of the lever, and β is the angle of the lever with the horizontal.

The electric motor rotates the lever at a rate of 3 degrees/sec. The door rotates freely ($W_k = 0$) until $t = 10$ seconds. The child clings to the door from $t = 10$ to $t = 20$ seconds. At $t = 20$ seconds, the child gets scared and lets go.

Write a MATLAB function, **PROB8_B_14**, that inputs the time the child first grabs the door and the time he eventually lets go. Your function should use a **FOR LOOP** to make subplots of the force applied by the lever versus time, and torque produced by the electric motor versus time from $t = 0$ seconds to $t = 60$ seconds.

Problem 8-B.15

The Bisection Method is a simple way of finding roots of polynomials numerically. Bisection Method starts with two guesses, x_1 and x_2, with the following relationships:

$$f(x_1) < 0$$

$$f(x_2) > 0$$

In such a case, the "zero value" of the polynomial (i.e., the root) must lie between x_1 and x_2. To close in on the root, we calculate the average of x_1 and x_2 and we evaluate the polynomial at x_3. According to the value of $f(x_3)$, we make one of the following decisions:

- If $f(x_3)$ is greater than zero, the root must lie between x_1 and x_3, so we replace x_2 with x_3.

- If $f(x_3)$ is less than zero, the root must lie between x_2 and x_3, so we replace x_1 with x_3.

- If $f(x_3)$ is zero, x_3 is a root, so we quit.

We continue the "average-replace" strategy until we find either a root, or that the difference between x_1 and x_2 is below a certain tolerance.

Use the Bisection Method described above in a function, **Prob8_B_15**, to find a root between $(x = 0)$ and $(x = 5)$ for the following polynomial. Set the tolerance in your solution to 10^{-3}.

$$f(x) = -x^4 + 4x^3 + 43x^2 - 58x - 240$$

Problem 8-B.16

A mass-spring system is set up as shown in the picture below:

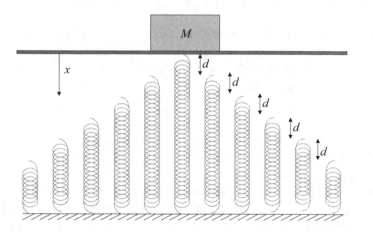

This arrangement shows 11 springs: one that touches the platform and 5 pairs of springs with d meters difference in heights. When mass M is supported by the platform, the platform will move through a distance x until the weight is balanced by the force of some number of the springs. The number of springs that it takes to balance the weight depends on the mass being supported by the platform (M).

A spring exerts a force proportional to its compression. The constant of proportionality is called the "spring constant." For a single spring of spring constant k supporting a mass of M, the balance of forces (gravitational vs. spring compressional) is as follows, given g as the acceleration due to gravity, 9.81 m/s^2:

$$Mg = kx \qquad if \ x \le d$$
$$or...x = \frac{Mg}{k}$$

If the mass being supported requires the center spring and the first spring pair out from the center to support it, then the balance of forces gives us the following:

$$Mg = kx + 2k(x - d) \qquad if \ d < x \le 2d$$
$$or...x = \frac{Mg + 2kd}{3k}$$

Write a MATLAB function, **Prob8_B_16_ver1**, that inputs M, d, and k and outputs the loaded position of the platform (x). Assume only three springs are available: the center spring and the spring pair closest the center spring.

Try your **Prob8_B_16_ver1** with inputs $M=10$ kg, $d=0.2$ m, and $k=200$ N/m. (Your answer should be $x=0.29667$ m.)

Now, generalize your solution to N pairs of springs, with spring pair $n=0$ to mean the single center spring and then count pairs of springs out from the center. When spring pair n is supporting the platform, the loaded position of the platform (x) can be calculated using the following:

$$x = \frac{Mg + n(n + 1)kd}{(2n + 1)k} \qquad if...nd < x \le (n + 1) \ d$$

Convince yourself that the above general relation is correct and be prepared to convince your instructor

Notice that you have to know which region the platform comes to rest in before you can compute its position. This seems circular since you have to know the position to compute the position.

But think of solving the general problem as follows: Assume the platform comes to rest in the region supported by only one spring, and compute the loaded position. If the loaded position turns out to be greater than the region of one spring only, then recompute with three springs. Continue until the loaded position that you calculate meets the assumption for the region the platform will be in when loaded.

Now, starting from your function **Prob8_B_16_ver1**, create a new function **Prob8_B_16_ver2** that takes inputs of N, M, d, and k and outputs the loaded position of the platform (x) and the number of the spring pair that is last to engage in supporting the platform. Understand what N stands for: the number of spring pairs in a system. Recall that $N=0$ represents the center spring.

Try your function with inputs you used before for version 1. The results for version 1 should agree with the results for version 2.

Now try inputs of $N = 11$, $M = 100$ kg, $d = 0.2$ m, $k = 200$ N/m. (Your answers should be $x = 1.9789$ m, and *springNumbe r* = 9.)

Problem 8-B.17

The following command line produces a random collection of lowercase alphabetical characters. Copy and paste this expression in the Command Window and observe its output.

```
char(97+floor(26*rand(1,10)))
```

Write a MATLAB program that displays a random collection of lowercase alphabetical characters and asks the user to type it. The program must do the following:

1. Display a text message that tells the user if he or she typed the string correctly. (Find information for comparing strings in MATLAB help.)

2. Time the typing speed of the user and display how long it took the user to type the string. (Search for **tic** and **toc** commands in MATLAB help.)

3. Ask the user if he or she wants to try again, and restart the game with a new string unless the user presses 'N' for No. The user should be able to play the game indefinitely by typing 'Y' at the end of each game.

Hint for using **input**: We do not want the user to put the string in single quotes; rather we want the user to press Y or N only. The command **input** accepts a second argument that tells MATLAB what type of input to expect, i.e., a number or a string. See the Help document for **input** to understand how to use its second argument properly.